Béla Gipp, Jöran Beel & Ivo Rössling:

ePassport: The World's New Electronic Passport

A Report about the ePassport's Benefits, Risks and its Security

www.ePassport-Book.com

Béla Gipp, Jöran Beel & Ivo Rössling:

ePassport: The World's New Electronic Passport

A Report about the ePassport's Benefits, Risks and its Security

www.ePassport-Book.com

Imprint

Jöran Beel
Zur Salzhaube 3
31832 Springe

Béla Gipp
Herzog-Wilhelm-Str. 63
38667 Bad Harzburg

Ivo Rössling
Olvenstedter Straße 61
39108 Magdeburg

Cover page photo by Boing (Photocase: `http://photocase.com`)

Acknowledgement

The following individuals and institutions deserve our special thanks for their contribution to this book.

Bundesamt für Sicherheit in der Informationstechnik (The German Federal Office for Information Security)

Michael Dickopf
Dr. Marian Margraf
Fabian Schelo

Bundesdruckerei GmbH (The German Federal Printing Office)

Dipl.-Ing. Ute Eberspächer

Bundesregierung (The German Federal Government)

Ulla Burchardt, SPD

EMPA, Zurich

Dipl.-Ing. Peter Jacob

Fraunhofer Institute, Berlin

Dipl.-Ing. Jan Krissler

Otto-von-Guericke University, Magdeburg

Dr. Martina Engelke

Unabhängiges Landeszentrum für Datenschutz (ULD) (Independent Centre for Privacy Protection (ICPP)), Schleswig-Holstein Schleswig-Holstein

Dr. Thilo Weichert

Thanks also go to:

Felix Alcala
Kirsty Brown
Bastian Czura
Stefanie Deichsel
Anja Gipp
Christian Hentschel
Birgit Lautenbach
Lars Petersen

Special thanks for proof-reading goes to:

Barbara Shahin

Preface by Henning Arendt

As owner of @bc® Arendt Business Consulting, a biometrics and security advisory, and project director of BioTrusT, the former IBM manager Dipl.-Ing. Henning Arendt, is intensively engaged in biometry and its suitability for the new passports.

Germany was one of the first countries to introduce the new electronic passport "ePassport" on 1st November 2005. The ePassport stores the facial image as well as the passport's reference data on a special chip integrated into the passport. This data can only be retrieved with a security procedure solely available to the authorities, when the passport is put on a special reading device.

Imagine this enormous challenge: throughout its whole period of validity of typically 10 years an issued ePassport shall allow biometric identification in a whole 189 countries (number of ICAO member states).

Since years I am actively contributing to national and international projects, aiming at reliable but also user-friendly identification based on biometric techniques. Already in 1999 my family began using biometric systems in everyday life: as entry permission to our house as well as access to information. Also from 1999 I was leading the perennial project BioTrusT. This project was supported by the German Federal Ministry of Economics, the savings bank organisation, and TeleTrusT. BioTrusT aimed at investigating all essential techniques w. r. t. their suitability for a broad usage in the banking field.

The BioTrusT project resulted in the nowadays international wide spread recommendations for the use of biometrics. This in particular covers the control of the biometric data by the user, as it has been realised now by German initiative for the ePassport. Everyone has his/her most individual biometric data stored only in his/her very own passport, rather than a central database.

Also thanks to German initiative, the encryption of the electronically stored reference data became internationally accepted and implemented in this first stage. During the first stage only the facial image is electronically retrievably stored on the ePassport. This is barely critical in comparison to the old non-electronically passports, since the bearer's photograph has already previously been part of any document of identity, visible for everyone being handed over the document.

The deployed encryption mechanism shall ensure that any read-out of this electronically stored image by unauthorised third parties will by all means be prevented. The next stage, in which also the bearer's fingerprints are stored electronically on the ePassport, demands far higher hurdles to be cleared in order to protect these reference data unique to any human being. In contrast to the facial image, the fingerprint reference data have not been recorded so far in the past.

This private data's merit for protection should be aware to everybody bearing responsibility for the ePassport. The economic damage would be tremendous, if unauthorised third parties were to obtain access to the reference data of certain persons and thereby take possession of their biometric identity. This of course especially applies to the reams of fingerprint recognition based entry and access systems in companies and authorities, that are currently used or in planning.

History has shown that only critical debates of new systems will result in a continuous improvement process. In the past, German industry was well capable of doing critical debates which resulted in Germany's industry being one of the world's best.

Therefore I recommend this book to all those who as mature citizens of any country introducing the ePassport, feel a need for reading up on this topic, but specifically to all those persons responsible for the next stages of the ePassport. Perhaps it will finally even be able to contribute to German security technologies, especially biometric solutions, becoming increasingly internationally accepted.

Hence, I am very pleased that the young and competent authors of this book provide you with this profound and critical insight into the relevant details. I would be happy to see the critical debate of this subject leading to more improved biometric systems solutions, achieving world-wide deployment and enabling comfortable and safe travelling.

Henning Arendt

Contents

List of Abbreviations

BAC Basic Access Control. The security mechanism designed for ensuring data privacy of the digital facial image.

BioPII Study of the BSI on the performance of biometric verification systems.

BKA Bundeskriminalamt. Germany's Federal Criminal Police Office.

BMI Bundesministerium des Innern. Germany's Federal Ministry of the Interior.

BSI Bundesamt für Sicherheit in der Informationstechnik. Germany's Federal Office for Information Security.

ECDSA Elliptic Curve Digital Signature Algorithm. The ciphering method used for the Extended Access Control.

ePassport (Electronic Passport). A passport featuring biometric traits such as face or fingerprint stored in the document in electronic form .

EAC Extended Access Control. The security mechanism designed for ensuring data privacy of the digital fingerprint.

FAR False Acceptance Rate. Percentage of invalid users who are incorrectly accepted as genuine users.

FRR False Rejection Rate. Percentage of valid users who are rejected as impostors.

FTE Failure to Enrol. Percentage of users for which the biometric trait cannot be enrolled successfully.

ICAO International Civil Aviation Organization. The organisation whose recommendations built the main basis for the development of the ePassport.

MRZ Machine Readable Zone. The part of the passport that is machine readable.

RFID Radio Frequency Identification. A method for contactless reading and writing a microchip, commonly used for the purpose of object identification. RFID systems generally consist of a transponder and a reading device, plus optionally an additional middleware and database.

RF(ID) Chip. Radio Frequency (Identification) Chip: The transponder which is used in an RFID system.

About the Authors

Jöran Beel[1], Béla Gipp[2], and Ivo Rössling are researchers at Otto-von-Guericke University Magdeburg, Germany, doing their PhDs in computer science and business. Their previous university studies focused on IT security, biometric application systems and theoretical computer science, respectively.

For their previous scientific work the authors received several honours. For instance, by the Heinz and Gisela Friedrichs Foundation for "extraordinary achievements in the field of technology development" and by former Federal Chancellor Gerhard Schröder for "outstanding scientific achievements". Following an invitation by Federal Minister for Education and Research, Edelgard Bulmahn, the authors presented their research results at the Hannover Messe 2003.

The authors also exhibit several years' professional experience and have cooperated with reputable firms such as Siemens and AOK. In addition, Béla Gipp works for the Working Group ISO/IEC JTC1/SC17/WG8, which is currently engaged in the standardisation of RFID technology.

---

[1]http://www.beel.org
[2]http://www.gipp.com

1 Introduction

With 1st November 2005, Germany was one of the first EU member states to introduce a new electronic passport. Compared to the previous non-electronic passport, this ePassport, named ePass in Germany, also contains the passport holder's facial image and from 1st November 2007 fingerprints, both stored in digital form. The ePassport's introduction is based upon a decision by the European Council. This decision stated that by the middle of 2006, all EU member states must issue electronic passports only.

Regarding its outward appearance, ePassports do not differ significantly from earlier passports. In addition to the paper part, ePassports contain a radio frequency chip (RFID-chip) that stores digital images of face and fingerprints. Moreover, the chip facilitates a way for contactless transmission of the stored data to a reading device in the course of border control procedure. In the follow-up, an automatic biometric recognition may take place in order to support the border control officer. During this procedure, the data stored in the passport is compared with the identifying person.

Basically, the usual identification procedure has changed only marginally. Previously, the border control officers manually performed a comparison of the passport photograph with the face of the assumed owner. Now, this is done by computer assistance. However, a comparison of biometric features, such as fingerprints, was not possible in the past in Germany.

Several data privacy activists and security experts have criticised the introduction of the ePassport. They apprehend the danger of unauthorized third parties unnoticed reading out the private data stored in the passports. Also, they fear that there may be unknown costs, and express uncertainty as to the performance of biometric recognition systems, wondering if they will eventually prove sufficient to ensure smooth functionality of the ePassport.

This book provides in six chapters a broad overview of the ePassport, its launch in Germany, and in particular, the raised criticism. Following this first introductive chapter, the subsequent Chapter 2 provides relevant information on the previous German non-electronic passport. Chapter 3 explains the ePassport and its basic functionality and the intended course of its introduction. Further technical details, with main focus on biometric features and security aspects, will be addressed in Chapter 4. Points of criticism, like possible ways to bypass security measures, reliability issues of biometric systems, or the durability of the ePassport itself, are discussed in Chapter 5. Chapter 6 gives a summary.

The information compiled in this book is mainly based upon the following documents:

[ICAO:2004a-i] The 1951 founded International Civil Aviation Organization (ICAO) has published a recommendation for travel documents augmented with biometric features. According to the European Council [EU:2004], this recommendation consists of several documents which constitutes the basis for the development and introduction of the ePassport into the European Union.

[BMI:2002a-b], [BMI:2005a-f], [BSI:2003], [BSI:2005a-e], [BSI:2006] Documents of the German Federal Ministry of Interior (BMI) and of the German Federal Office for Information Security (BSI) provide users basic and relevant information to the ePassport and its introduction in Germany.

[BioPII:2005] In cooperation with the German Federal Criminal Police Office (BKA) and secunet GmbH, the German Federal Office for Information Security (BSI) conducted a large-scale "Trial on the Performance of Biometric Verification Systems – BioP II". The field study was aimed at investigating the suitability of biometric features; face, fingerprint and iris, with respect to their use within electronic passports. Thereby, the main focus was on examining to what extent reliable verification proves feasible based upon quoted biometric features. Apart from that, acceptance and ease-of-use were considered. Although the test population may not be regarded as representative for the overall German population [BioPII:2005, p. 56], the study still served as a good starting point, as the guidelines of the ICAO were fundamentally incorporated into the study as far as general conditions were considered. [BioPII:2005, p. 10]. Altogether, about 2.000 individuals participated in this study [BioPII:2005, p. 51].

[UKPS:2005] From April to December 2004, the UK Passport Service (UKPS), in cooperation with the Home Office Identity Cards Programme and the Driver and Vehicle Licensing Agency (DVLA), conducted a study with a total of 10.000 participants. The "Biometric Enrolment Trail" was aimed at examining the maturity of biometric systems, putting the main focus on enrolment and verification. In addition, the study wondered whether and to what extent the systems are suitable for disabled persons. The study may not be considered fully representative for the overall UK population [UKPS:2005, p. 8].

These documents are cited at the respective locations, as well as further relevant resources, all of which listed in the bibliography in the back of this book.

Unfortunately, this book has been published without approved illustrations of the ePassport. All rights regarding existing illustrations are reserved by the German Federal Ministry of Interior, which in turn explicitly prohibits the use of any such illustrations within this book.

2 The Former (German) Non-Electronic Passport

2.1 Introduction

To evaluate the need for introducing a new electronic passport, it is necessary to figure out to what extent the non-electronic passport meets today's requirements. In this respect, data security and data privacy of the former (German) non-electronic passport are the two aspects that this chapter deals with.

After a brief introduction which provides general information on the former passport, the focus lies on the data security provided by the former German passport, followed by an examination of the data privacy. The summary in the final section of this chapter provides an assessment on the suitability of the former passport and the necessity for a revised version.

2.2 A Brief Survey

More than 65 Mio copies of the former non-electronic German passport have been produced by the Federal Printing Office[3] since its introduction in 1988 [BMI:2005a]. The document was granted a 10 year duration of validity. If the applicant has not reached the 26th day of birth, this validity was limited to 5 years [AA:2005a].

Personal data about the passport holder were captured in both machine-readable format (i. e. OCR) and human-readable format. The so-called preliminary passport, with a one-year validity, was not machine-readable in the past, but now is since 1st January 2006 [AA:2005b]. However, not all public authorities are yet able to issue these documents [AA:2005a].

Apart from signature and photograph of the passport holder, the following information was stored for the purpose of identification [PassG:1986, §4]:

- Surname, First Name, Middle Initial
- Any other given names (maiden name)
- Doctor's Degree (if applicable)
- Religious name or pseudonym

---

[3]http://www.bundesdruckerei.de

- Date and place of birth
- Sex
- Height
- Colour of eyes
- Residence
- Nationality

2.3 Data Security

For any state, the aspect of data security plays an important role. The passport data, and of course the document itself, shall not be allowed to be manipulated, counterfeited or forged at all – or at least with very high difficulty only. This is the only way to counteract the risk of passport abuse, e. g. for entrance into the country under false identity.

In an interview with the German magazine SPIEGEL, Germany's former Minister of the Interior, Otto Schily, who was instrumental in the introduction of the ePassport, stated the presumptive existence of a series of forgeries of German identity cards and passports [Spiegel:2001]. However, he did not mention any specific numbers. In general, the German non-electronic passport was said to be one of the most secure and most fraud-resistant travel documents worldwide [BMI:2005a, BMI:2005b], especially since 2001. In November 2001, the German passport added eight additional and completely newly developed security features, which have been compiled by the Federal Printing Office and the Federal Criminal Police Office [BMI:2002a, BDR:2005b].

These new features were:

1. Holographic portrait
2. 3D German eagle
3. Kinematic structures
4. Macro lettering and micro lettering
5. Contrast reversal
6. Holographic representation of the machine-readable lines
7. Machine-verifiable structure
8. Surface embossing

Three more security features were adopted from the first version of the passport:

9. Security printing with multi-colour guilloches

10. Laser lettering

11. Watermark

In 2002, the Federal Police Central Bureau investigated 7700 passports [Bund:2005]. Among these, 290 were total forgeries from EU countries (93 from non-EU countries), 394 passports were content-based modified original documents from EU countries (1086 from non-EU countries), and in 91 cases, it was an issue of purloined blank passports (with the latter ones not being distinguished between EU or non-EU countries). Among those forged passports, 35 were German. However, most of them were not forgeries of regular passports, but instead of preliminary ones (specific numbers are missing in the report).

According to the Federal Criminal Police Office, the focal point of forged passports is located in the following countries: Italy, France, Spain, Greece, Portugal and Belgium [BMI:2005b]. This data, however, only relies upon information provided by the Federal Police Central Bureau. As stated by Otto Schily, a worthy number of further counterfeits had been seized by the German prosecution agencies [Spiegel:2001] – among these are also forgeries of German passports. It stands to reason that at least at the European level, there is room for improvement concerning forgery-proofness.

2.4 Data Privacy

The federal passport law of Germany [PassG:1986] and the registration acts of the respective federal states commit Germany's citizens to provide their personal data to the competent registry office.

According to the intention of the Federal Data Protection Act, the personal data shall be acquired directly from the persons concerned – requiring their physical presence and acknowledgement [BDSG:2007, §4]. Additionally, the persons concerned are to be given the opportunity to decide without interference, which personal data they want to provide and to what extent usage will be granted [BVerfGE:1983, p. 43 et sqq.]. This directive expressly includes that the personal data must not be raised unnoticed, nor read out or misused by third parties [ULDSH:2003].

To ensure data privacy with respect to the passport, the federal passport law [PassG:1986, §16] states that personal data of the passport holder may only be stored by the competent registry office and within the passport itself. Any storage somewhere else is inadmissible. In particular, the "Bundesdruckerei GmbH" (– the Federal Printing Office, manufacturer of the passport –) is bound by law to erase any personal data immediately after manufacture of the document. However, the Federal Printing Office is granted to maintain a database about all the unique serial numbers of the passports ever issued, although for the limited purpose of proof of disposition only [PassG:1986, §16]. Besides the Federal Printing office, this database may additionally be accessed by the competent passport authority, the Federal Police and the police authorities of the individual Federal States. (For details, see [PassG:1986, §16]).

Unauthorised access by third parties appears to be almost impossible. The passport holders keep control of their personal data as long as they do not hand over the document to unauthorised persons, the passport does not get lost, and all actual entitled authorities (see above) abide the law.

2.5 Conclusion

The former German passport was one of the most fraud-resistant passports worldwide. Data privacy was widely warranted. However, forgeries of other, also European countries' passports occurred rather frequently. The call for increasing the security on European level appears basically justified.

3 The ePassport – A Survey on the New Electronic Passport

3.1 Introduction

To strengthen the binding of the passport to its unique holder and to make it more fraud resistant, the EU member states and some other countries like Australia, the United States, Japan, Russia and Switzerland have collectively decided to introduce a new type of passport - the ePassport. In Germany, this newly-named *ePass* has already been launched since November 2005.

The ePassport is equipped with an electronic part for storing biometric data. The increased forgery-proofness and document binding shall provide an effective remedy against terrorism, organised crime, illegal immigrants and identity theft. This chapter gives an overview about the basics of the ePassport, its aims and application in practice. Finally, a conclusion will be given.

3.2 The Basics

On 13th December 2004, the European Council resolved the introduction of a new passport – mandatory for all EU member states [EU:2004]. The crucial point of this regulation was the new obligation of the electronic storage of biometric data within passports – more precisely, the data on face and fingerprints of the passport bearer. Thereby, the EU resolution on the introduction of electronic passports adheres closely to the recommendations published by the International Civil Aviation Organization (ICAO)[4]. In this organisation, Germany is represented by the Ministry of the Interior, with technical assistance of the Federal Office for Information Security (BSI) and the Federal Criminal Police Office (BKA) [BMI:2005d].

While the ICAO recommends that it be mandatory to store the facial image and other biometric features such as fingerprints or iris data be optional [ICAO:2004e, p. 15], the EU ordinance demanded all the member states from the middle of 2006 to store the facial image, and from 2008, the fingerprints [EU:2004]. The BMI explained the decision of the EU as follows:

"The commitment of the European Union to use two biometric features was necessary in order to provide additional flexibility to the passport control process. In situations where facial recognition turns

---

[4]http://www.icao.org

out to be impractical (e. g. in the evidence of poor illumination or haste of people), a verification based on fingerprints should be performed[5] [BMI:2005e]

Germany already launched the first version of the ePassport, thus earlier than specified by the EU ordinance. On 8th July 2005, the Federal Council of Germany finally approved the decision that was taken by the Federal Government to issue electronic passports from 1st November 2005 [BR:2005].

In the initial step now, the chip integrated in these passports is storing an electronic version of the bearer's facial image. This image is to be identical with the photo printed on the data sheet [ICAO:2004d, p. 33]. Originally it was intended to store two fingerprints on the ePassport as of March 2007 [BMI:2005c]. But the date was re-scheduled to November 2007 [BMI:2007]. The fingerprints will not be printed on the paper part of the ePassport.

Old passports without any biometric features and first generation ePassports storing only an electronic version of the facial image will keep their ten-year validity, even though passports issued from November 2007 are solely issued with additional electronic fingerprints [EU:2004, Art. 6]. The necessary changes to the law allowing to incorporate biometric features into German passports were already legislated in 2002 in the course of the Anti-Terror Act [BMI:2002b]. All border controls must be fully equipped with appropriate scanners, and this has recently been started and it is estimated that it shall be finished by 2008 [BSI:2005a].

According to an EU resolution [EU:2004], each EU member state may hold only one single authority to issue the electronic passports. In Germany, this authority is the Bundesdruckerei GmbH, the Federal Printing Office. The Philips AG and the Infineon Technologies AG, in this process, supply the RFID-chips meant for storing the biometric data [BMI:2005c]. Both the BSI and the BKA provide support to the Federal Printing Office in developing new security standards [BMI:2005d]. The last part of the process includes the Flexsecure AG, which is responsible for providing the software for operating the Country Signing Certification Authority (CSCA). There is still severe criticism from several sides concerning the launch of the ePassport – especially with respect to the way the decision-making process was carried out at both the European and federal levels.

Chapter 5.6.3 deals with this subject and addresses the most common points of criticism.

---

[5]Translated from German.

3.3   Aims of the ePassport

By storing the facial image and fingerprints digitally, the bond between the passport and its unique holder shall be strengthened [BSI:2005c]. It shall increase the reliability when deciding whether a given person is actually the one being identified by the document. Together with digital signatures, this also shall make the ePassport more forgery-proof [BSI:2005c]. With increased document binding and fraud resistance it is going to be much harder to take possession of a false identity, e. g. in order to enter the German federal territory illegally.

According to the BMI, the ePassport thus embodies an effective instrument against terrorism, organised crime, illegal immigrants, and identify theft [BMI:2002b]. Since the ePassport is being introduced across Europe, the gain in security for Germany even multiplies, because so far, predominantly passports other European countries have been forged [BMI:2005c]. In addition, "trustworthy" persons might benefit from the new ePassport by means of a considerable amount of saved time during border control checks [BMI:2005b].

Germany's former Minister of the Interior, Otto Schily, mentioned a refreshment of the German economy and the unique chance to demonstrate "that Germany has the necessary know-how and innovation to set standards in the emerging sector of biometry" [5] [BMI:2005c]. Moreover, a study of the European Commission states "A successful appliance of biometry in the ePassport will lead to reduction of anxieties and prejudices of the public concerning biometry in general [EU:2005]." This way, biometry would become accepted much easier in other fields of life as well.

It appears plausible that the same will also hold for the efforts towards the use of RFID, which will serve for both the storage of media and data communication in the ePassport. It must be said, though, that substantial problems during the introduction of the ePassport might, in turn likely result in the opposite, an intensification of the reservations concerning biometry and RFID, and thus a loss for the economy.

The overwhelming question to what extent a broadening of biometry and/or RFID in areas other than identification documents is desirable at all shall not be part of this book. Whether the objectives which Otto Schily claimed to be "of overall German public interest" [6] [BMI:2005b], will actually be achieved with the introduction of the ePassport, is questioned by some critics (cf. Chapter 5).

---

[6]Translated from German.

3.4 The ePassport in Practice

The German ePassport includes all the security features known from the old passport. Additionally, it contains a Radio Frequency Identification Chip (RFID-chip). The RFID-chip serves as a storage media and for transmitting the facial image and fingerprints of the passport holder [BSI:2005a]. Furthermore, the chip stores the hitherto existing personal data, such as residence, date of birth, etc. [BSI:2005b].

With respect to their functionality, RFID-chips can be compared to the chips on money or phone cards – except that they can be read out contactlessly.

In accordance with the EU resolution [EU:2004], the passport holder can verify at any time in detail what data actually is stored on the chip, using scanners located in the registration offices.

As for the populace, little change took place with the launch of the ePassport on 1st November 2005. The procedure for applying for a new passport stays nearly the same. The only difference is that the passport photograph will henceforth need to be a full-frontal shot. This change is actually based on ICAO guidelines [ICAO:2004c] and the fact that the automated face recognition performs best if based on full-frontal shots of the facial images [Bioface:2003].

The applicant has to provide the passport image in standard size, printed at a resolution of at least 600dpi to the registration office. A submission in digital format is not permitted because of the possibilities of the risk of manipulation. The staff of the passport authority checks if the photo conforms to a few basic guidelines, which are provided to the applicants in advance, in form of e. g. posters at the registration offices. However, an explicit (and possibly computer-assisted) in-depth analysis of the passport photo regarding the provision of adequate biometric features will not take place, nor does a complete enrolment [Heise:2005e].

As with the former passport, the German ePassport is granted a ten-year validity period [BSI:2005a]. The costs currently amount to 59,– Euro per exemplar – or 37,50 Euro for a passport with a five-year validity (for passport holders younger than 26). Costs for ePassports in other countries vary.

According to the EU resolution [EU:2004], short-term passports with less than a one-year validity do not necessarily need to be equipped with an RFID-chip nor with biometric data. The same holds for identity cards. However, as of 2008, Germany will also issue only biometric-based identity cards [Bund:2005].

In the course of the border control check, the traveller will have to face very few changes, in the ideal case. Scrutiny is still carried out by border police officers. The biometric features are only to support them with their work [Bund:2005]. Actually, even an ePassport with a defective RFID-chip will not get invalidated [BSI:2005a], as well as any of those documents that were issued before the launch of the ePassport, will keep their validity (until their regular expiration) [EU:2004, Art. 6]. In the latter cases however, the person may expect a more intensive security check [CCC:2005a].

Figure 1 shows the operational sequence of the passport inspection at the border control.

For the protection of the sensible personal data stored on the RFID-chip against an unauthorised readout or unauthorised monitoring of facial image or fingerprints (if present), the ICAO has defined two relevant standards [ICAO:2004a]: *Basic Access Control* and *Extended Access Control*.

In short, *Basic Access Control* ensures that, as a rule, the biometric data may only be read out after an optical scan of the MRZ.[7] The data communication is carried out in encrypted manner, for the sake of rendering unauthorised eavesdropping nearly impossible. The extended protection for accessing the biometric data, the *Extended Access Control*, is based on public-key encryption intended for preventing the personal data from being read out by third parties. Further details concerning the access protection will follow in the next chapter.

Moreover, the suggestion of the ICAO governs that only the facial image, which is obligatory, has to be accessible for all countries in the context of passport control. Additional biometric features may be subject to approval by selected countries, based on encryption and certificates. For those countries to which Germany will grant access to the fingerprint, seems as of yet, not determined [CCC:2005a]. Problems resulting from this will be discussed in Chapter 5.6.5.

3.5   Summary

On 1st November 2005 the new electronic passport, the ePassport, was launched in Germany. Initially, this passport stored a facial image of the document holder on its embedded RFID-chip. From

---

[7]The machine readable zone (MRZ) is a special area in the passport, providing the document owner's name, date of birth and gender in some particular machine-readable format, allowing data extraction via OCR.

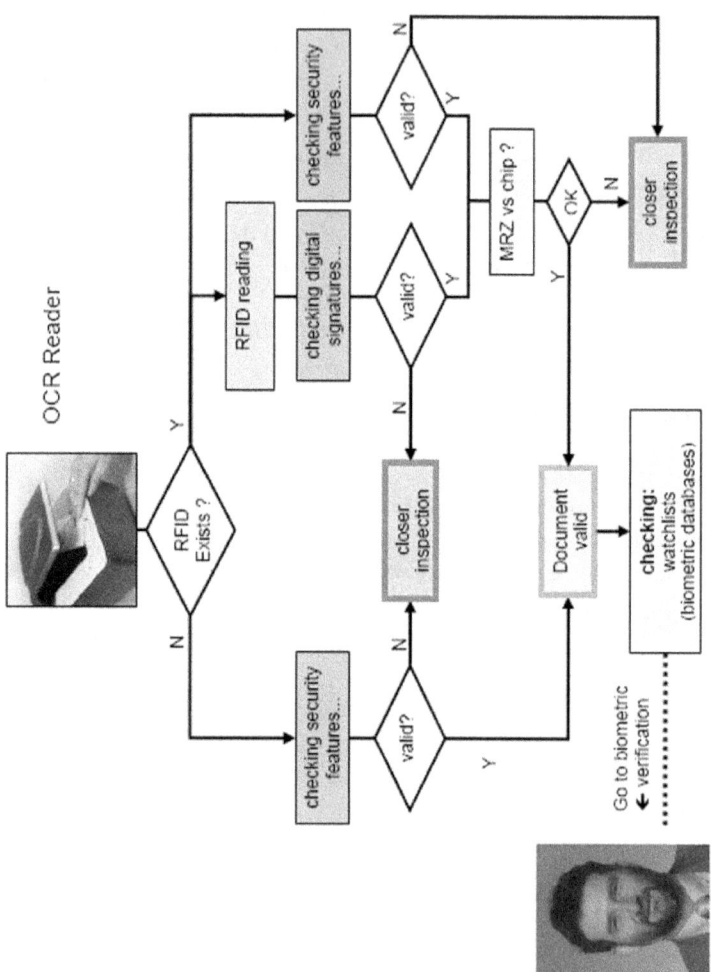

Figure 1: Operational sequence of the passport inspection (Source: [ICAO:2004d, p. 44])

November 2007 on, two fingerprints will additionally be included. The price for the German ePassport has been raised from 26,– Euro to 59,– Euro. Inasmuch as the roll-out and operation of the ePassport continues to go according to plan, the passport owner will hardly face any changes. The most common points of criticism will be addressed in Chapter 5.

4 The ePassport in Detail – On Technical Specifications and Biometry

4.1 Introduction

Besides the paper sheets, the ePassport contains an RFID-chip. This chip is meant for storing the biometric data and serves for both data transmission and encryption. This chapter provides a survey about the basic technical specifications of the RFID-chip, gives an overview about the selected biometric features, and assesses their suitability for being used for the ePassport. Afterwards, a survey on security measures for protecting the personal data will be given. The chapter finishes with a summary.

4.2 RFID

Basically, an RFID-system consists of two components: a transponder and a reader [RFID:2002]. In the case of the ePassport, the transponder is the radio frequency chip (RFID-chip) embedded in the document body. The chip's EEPROM holds the biometric data and the microprocessor provides the functionality to ensure the security functions described in the following sections.

In its recommendation, the ICAO argued that RFID-chips for electronic passports should be compliant with the ISO/IEC 14443 standard [ICAO:2004b]. These chips operate with a frequency of 13.56 MHz, and according to the ICAO, provide several advantages over competing standards and technologies. The frequency band, for example, is deemed to be serviceable in every country world-wide. Furthermore, data transmission on this frequency cannot be disturbed by water or the human body, but on the other hand, could be shielded by the use of metal. Chips compliant to this standard have already been in use for a couple of years and are tried and tested. Moreover, they provide ample memory capacity, a sufficiently high data transfer rate and afford the opportunity to read out multiple passports simultaneously. Finally, the design of the chip permits it to be embedded into passports without altering their essential format. Moreover, the chips are powerful enough to perform the necessary encryption and identification functions. The restricted read-out range (typically up to 10cm), makes unauthorised reading-out and signal sniffing more difficult compared to other standards, where frequencies and transmission powers permit reading-out within a distance of up to several meters [ICAO:2004b].

The reading device at the border crossing is able to supply the RFID-chip via induction with electricity and read out the data contactless. This book, however, will skip further details on the respective physical basics. Instead, the reader may refer to [RFID:2002] for more detailed information. The important fact is that communication between the reading device and the transponder (ePassport) is contactless and that the ePassport does not require a separate power supply (batteries, generators, etc.).

Since data is transmitted contactless, there is a fundamental risk for transmissions (and thus data) getting recorded by non-authorised third parties. To prevent data from being sniffed, encryption is used (cf. Chapter 4.4). Regarding the security of the encryption method used therein, criticism is still being heard from several sides (cf. Chapter 5.5).

Germany's former Data Protection and Information Freedom Officer (BfDI), Peter Schaar, for instance, called for a 3D bar code to be used instead of the RFID technology in order to satisfy data protection requirements [Heise:2005a]. Table 1 compares the most common alternatives.

Within its recommendation, the ICAO has dealt with several alternatives, but has finally found the RFID standard ISO 14443 being the most appropriate.

The deciding factor, according to [ICAO:2004d, p. 35], was the given fact of RFID having been the only technology to meet the demands with respect to usability, data capacity and performance.

Indeed, contact-based chip cards could deliver the required functionalities. However, the ICAO casted doubt on a contact-based chip being still fully functional after 10 years, since it would most likely show wear and tear, and the contacts for instance could have been damaged by oxidation as well. Moreover, the embedding of such chips turned out to be difficult, if not impossible within passports of traditional design.

4.3 Biometry

4.3.1 Introduction

Besides the increased need for a forgery-proof medium, the main reason for the introduction of the ePassport was due to the biometric features stored in the passport allowing a more reliable assessment about persons identifying themselves as being the legal passport owner [BSI:2005c].

Parameter/System	Bar Code	OCR	Smart Card	RFID
Typical Data Volume [Byte]	1~100	1~100	16k~64k	16k~64k
Data Density	small	small	very high	very high
Machine Readability	good	good	good	good
Human Readability	limited	easy	impossible	impossible
Influence of Dirt / Moisture	very high	very high	possible	none
Influence of (opt.) Coverage	total breakdown	total breakdown	possible	none
Influence of Alignment and Positioning	little	little	very high (plug-in connector)	none
Wear and Tear	limited	limited	limited	none
Acquisition Cost Readout Electronic	very low	medium	low	medium
Unauthorised Copying/ Altering	easy	easy	hard	hard
Readout Speed (incl. Handling of Storage Medium)	slow (~4s)	slow (~3s)	slow (~4s)	very fast (~0,5s)
max. Distance between Storage Medium and Scanner	0-50cm	<1cm	direct contact	0-5m

Table 1: Properties of selected Auto-ID systems by comparison (Source: [BSI:2004a, p. 90])

This section presents biometric techniques and comments on advantages and disadvantages of their use in the ePassport.

The term "Biometry" derives from the Greek words "Bios" (the life) and "Metron" (the measure), cp. [Duden:2005]. Biometric features of a human being refer to measurable, preferably individual features of the body that ideally change only marginally throughout lifetime.

Besides biometric techniques, there are two other ways of authenticating an individual. The identity of a person could on the one hand be verified by specific secret knowledge, for instance a password, or by personal possession of a special item, e. g. a car key, own to the individual.

In both cases, however, no personal bond is actually being ensured, which in turn makes for fraudulent use and is thus unsuitable for the object of verifying the legitimate holder of a document of identification [BSI:2005a].

4.3.2 Overview on Biometric Techniques

Not every feature is equally suitable for the object of biometric authentication. In fact, the following criteria should be fulfilled [Bromba:2005a, Bromba:2005b]:

Acceptance: The method should be accepted by the users and must not harm or compromise the dignity or health of human beings.

Permanence: The biometric feature should not alter over time beyond a certain tolerance.

Disponibility: The biometric feature should be present for every user

Measurability: The biometric feature should be measurable with reasonable technical effort.

Uniqueness: Not every biometric feature is unique. Most appropriate are features that develop during the (pre-)embryonic stage based on random processes during tissue growth (randotypical). Examples are the individual structures of irises or fingerprints, which even differ between monozygotic twins. Rather unsuitable are behavioural characteristics, since they are either learnable or genotypic, thus hereditary. [BSI:2005a]

Table 2 provides an overview of biometric features that basically satisfy the criteria mentioned above.

Based on the technical report of the New Technology Working Group (NTWG), the ICAO considered the three features face, fingerprint and iris as being the most suitable ones for the use within documents of identification [ICAO:2004d]. They suggested the digital photo of the face to be mandatory and other biometric features to be optional [ICAO:2004c].

However, the EU albeit passed a resolution binding to all member states to not only record the facial image, but additionally the fingerprints [EU:2004]. An incorporation of iris data was not mentioned in this resolution.

Biometric feature	Description
Fingerprint	Finger lines, pore structure
Signature (dynamic)	Writing with pressure and speed differentials
Facial geometry	Distance of specific facial features (eyes, nose, mouth)
Iris	Iris pattern
Retina	Eye background (pattern of the vein structure)
Hand geometry	Measurement of fingers and palm
Finger geometry	Finger measurement
Backhand vein structure	Vein structure of the backhand
Ear form	Dimensions of the visible ear
Odor	Tone or timbre
DNA	DNA code as the carrier of human hereditary
Smell	Chemical composition of the one's odor
Keyboard strokes	Rhythm of keyboard strokes (PC or other keyboard)

Table 2: The most well known biometric features used for authentication purposes (Source: http://www.bromba.com)

The German Federal Government at present refrains from incorporating iris data into the ePassport [Bund:2005]. However, according to a report of the c't magazine [Heise:2005f], Germany's former Minister of the Interior Otto Schily supports the iris data to be added to the ePassport in the long run. One of his speeches gives rise to the presumption that he considers the iris to be especially suitable for documents of identification [BMI:2005b].

In the following sections, the three biometric features face, fingerprint and iris will be examined with regard to their suitability for being used in the ePassport.

The one thing that all biometric features have in common is, that the effect of aging upon recognition performance hasn't been investigated sufficiently until now [BioPII:2005, p. 170]. Hence, it is hard to assess, how well the recognition performance will be in ten years, based on features that have been recorded today.

Furthermore, for all three features, the recognition performance based on reference images (as recommended by the ICAO) at present still yields 1–2.5 percentage point less than those based on using templates[8] [BioPII:2005, p. 165]. The use of raw data[9] in turn, provides the advantage that one can abstain from proprietary templates that potentially might not be compatible with all kinds of reading devices [ICAO:2004d].

Throughout the following sections, the terms FAR, FRR and FTE will be used. They convey the following meaning:

False Acceptance Rate (FAR): The False Acceptance Rate is defined as the ratio that an unauthorised user will incorrectly be accepted by the biometric system, although the verification should have failed, since the biometric features of the person disagree with the reference values. [ICAO:2004d, p. 10].

False Rejection Rate (FRR): The False Rejection Rate is defined as the ratio that an unauthorised user will be rejected by the biometric system, although the verification should have succeeded [ICAO:2004d, p. 10]

Failure To Enrol Rate (FTE): The Failure To Enrol Rate is defined as the ratio of unsuccessful enrolments due to not clearly defined or non-existent biometric features [SchimKVK05].

4.3.3 Face Recognition

Since November 2005, an electronic version of the facial image is stored in German passports (cf. Chapter 3). The ICAO sees several advantages in the use of the facial image as a biometric feature [ICAO:2004d, p. 17]. Facial images, for example, do not uncover any information that would not be revealed already from simply showing your face in public. Facial images are internationally accepted for use in documents of identification. Between the user's face and the reading device, no contact or direct interaction is required. Moreover, no special equipment is needed for the enrolment of this feature.

Another fact speaking in favour of face recognition is that the feature "face" is present on virtually every human being and can, in general, easily be enroled [BioPII:2005, p. 12] & [UKPS:2005, p. 9]. However, in some rare cases the enrolment might still fail. In one study, one participant featured such a dark skin colour that even

---

[8]Proprietary templates are manufactorer specific data formats incorporating only those data relevant for the actual recognition and thus involving only a fraction of the storage capacity demands of raw data.

[9]Raw data are untreated original data; for instance, the facial image in JPEG format.

after seven attempts, no successful enrolment could be achieved [UKPS:2005, p. 196]. The system consequently classified his skin as "underexposed area".

One fact against face recognition is that simple marginal changes in lightning can lead to drastic changes in recognition performance [BioPII:2005, p. 16]. This disadvantage also occurs in the context of 3D face recognition. The BSI runs tests on their praxis capability [BSI:2005d]. However, since 3D face recognition is neither mentioned in the ICAO documents nor in the EU resolution, it will not be examined any further in this book.

Benchmark values on the performance for 2D face recognition vary a lot, depending on the particular study. While the BioFace study from 2003 states a FRR of 50% [Bioface:2003, p. 10], the BioPII study from 2005 reports on a FRR of 2–10% with a FAR of 0,1% [BioPII:2005, p. 12]. One explanation for this discrepancy might surely be the progress in the technical development in the last two years. However, a study by British passport authorities, also from 2005, states a FRR of about 31%, whereas the FAR is not specified [UKPS:2005, p. 10].

The influence of a user's occupation on recognition performance is unclear [BioPII:2005, p. 127]. The influence of gender, in turn, seems rather to be less determined by biometry, but more by the used reading devices. BioPII determined a gender dependency for some of the devices, leading to a slightly higher FRR for female users [BioPII:2005, p. 126]. For other devices, in turn, no dependency could be observed. In [UKPS:2005, p. 240] as well, no dependency could be observed. The influence of ethnical origin is unclear [UKPS:2005, p. 239], a link however is assumed [Bioface:2003].

Age seems to play a significant role with regard to recognition performance. [UKPS:2005, p. 241] states that the recognition performance decreases noticeably for individuals over sixty years of age. [BioPII:2005, p. 127] notes that the recognition performance is likely to be dependent on the age factor, but since the number of participants over sixty was only at about 2%, this statement should be considered carefully [BioPII:2005, p. 53].

4.3.4 Fingerprint Recognition

As from November 2007, two fingerprints will be incorporated into the German ePassport (cf. Chapter 3). The BSI study "BioFinger" [Biofinger:2004] subdivides the process of fingerprint analysis into six stages.

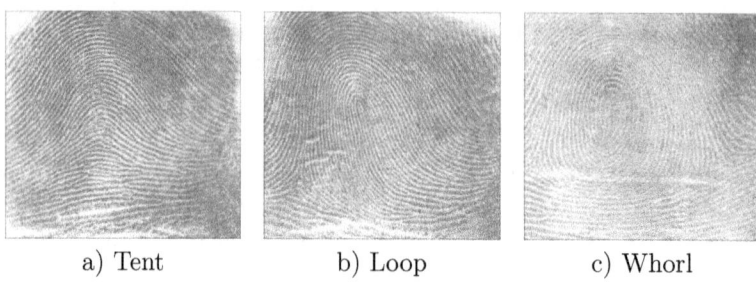

a) Tent   b) Loop   c) Whorl

Figure 2: Coarse features of fingerprints (Source: [Watson:2005])

First of all, the pass holder needs to be fingerprinted. This can be done by taking an impression on paper or metal plates or by using an electronic sensor. As for the sensors, there are different types available, a.o. capacitive, optical, pressure sensitive and thermal ones, each of them featuring different advantages and disadvantages [Bromba:2005b].

After having a digital version of the fingerprint at hand, several image processing algorithms are applied to enhance image quality followed by an image-editing process.

During the stage of pattern classification, several coarse features of basic type such as loops, arches, whorls (cf. Figure 2), etc. are identified.

In the phase of feature extraction, the fingerprint's fine features, the so-called minutiae, are located, resulting from the existence of line branchings and endings. The relative position of these minutiae to another, as well as their type makes fingerprints unique and comparable for the algorithms. The quantitative factor specifies the number of minutiae found.

During the final verification phase, the degree of correlation is determined, and according to the specified threshold, the fingerprint is classified as identical or non-identical.

In both [UKPS:2005, p. 221] as well as [BioPII:2005, p. 12], in 1% of the participants, the fingerprints could not be enroled successfully. [BioPII:2005] assesses recognition performance to be very good, with a FRR of 1–7% and a FAR of only 0,1%. An older study from 2004 arrived at the same conclusion with the FRR to be at about 2% and the FAR of 0,1% [Biofinger:2004, p. 3]. On the contrary, [UKPS:2005] reported an FRR of 20%, unfortunately without specifying the FAR.

In any case, recognition performance depends on the age of the user, leading to a slightly lower FRR for younger people and a higher rate for older ones [UKPS:2005, p. 249] & [BioPII:2005, p. 127]. Moreover, the aging process of this biometric feature itself seems to be relatively high. The FRR for instance is likely to double in the case of a comparison with a fingerprint recorded 10 years ago [Biofinger:2004, p. 3].

Furthermore, recognition performance is presumed to be dependent on gender, too. Male users achieve better results due to their larger hands [UKPS:2005, p. 251] & [NIST:2002, p. 7]. If recognition performance is additionally dependent on ethnical origin, it could not be stated definitely, although it seems likely [UKPS:2005, p. 247].

4.3.5 Iris Recognition

At present, an incorporation of iris data into the German ePassport is not planned specifically. However, the feature is nevertheless being considered at this point, since a later incorporation appears to be likely (cf. Chapter 4.3.2).

The iris structure is suitable as a biometric feature, since it is – like the fingerprint – randotypical, i. e. developed at the (pre-)embryonic stage, based on random processes and thus even different for monozygotic twins. Another advantage is the complexity of the structure. During an iris-scan, about 250 unique features can be identified, which in turn leads to a theoretical probability of $1 : 10^{78}$ to compute the same iris template for two different persons [GES:2005]. Fingerprints, however, have only about 50 features that are identifiable, depending on the quality of the image and the degree of distinctness of the values [WDR:2005].

Another advantage of iris recognition is the possibility of contactless acquiring which is desirable for hygienic reasons.

[UKPS:2005, p. 10] states the FRR of iris recognition to be at about 4%, while [BioPII:2005, p. 164] reports the FRR between 2% and 25% with a FAR of 0,1%. The high variation of the BioPII study is caused by the distinction of frequent users and infrequent users within the context of the statements.

While users that were frequently identified via their iris (more than 120 actuations) showed a lower false rejection rate, users with rare contact (less than 10 actuations) got more frequently false rejections[BioPII:2005, p. 12+167].

This aligns with the statement of the BioPII study that iris acquisition and recognition systems still show a need for improvement with respect to their usability. It can be expected, so the study says, that especially for infrequent users the recognition performance will indeed significantly improve as soon as the usability of the systems gets enhanced [BioPII:2005, p. 13].

The poor usability also explains the very high Failure To Enrol Rate of about 10%, reported in [UKPS:2005, p. 9]. The acquisition devices sometimes did not return any (appropriate) feedback in case of a failure, and users with extremely poor eyesight were unable to locate a certain point whose fixating was fundamental for a successful enrolment [BioPII:2005, p. 94] & [UKPS:2005]. Moreover, users with a body height of less than 1,55m could not be enroled at all, or with great difficulty only. Due to the design of the test systems, the camera was unable to acquire the iris for these candidates [BioPII:2005, p. 94].

According to [UKPS:2005, p. 245], recognition performance seems to be gender-independent. A dependency on age could in turn be observed. For users under 20 or over 50, recognition performance abates significantly [UKPS:2005, p. 244] & [BioPII:2005, p. 127]. Furthermore, [BioPII:2005, p. 128] reports that recognition performance is lower for blue collar workers than for white collar workers ones.

The most important matter raised against the use of the iris as a biometric feature is the problem of acceptance [BioPII:2005] & [UKPS:2005]. Theoretically, an iris-scan might permit an acquisition of health data. It is unclear though, to what extent data attained that way are utilisable for scientifically founded diagnoses [AOK:2005], [Aetna:2005].

4.3.6 On Storing Biometric Data

Normally for the purpose of biometric recognition, so-called templates are used [BioPII:2005]. They are generated by extracting relevant features from biometric raw data. At present, though, there are no general standards for template generation. Hence, manufacturers of reading devices adopt different templates, which are largely incompatible to each other. In order to assure that the electronic passport can be processed by different reading devices, the ICAO recommends the use of raw data instead of proprietary templates [ICAO:2004a, p. 17]. For national purpose then, templates might still be stored additionally on the RFID-chip.

Since ordinary digital images may come along with high storage requirements, the ICAO conducted studies in order to analyse the relationship between image size and the performance of biometric recognition [ICAO:2004f] & [ICAO:2004g]. Figure 3 shows the performance of face recognition based on JPEG-compressed images as a function of the image size.

The ICAO concludes that a facial image of a 420 pixel width and a 525 pixel height in JPEG-format with a file size of 15–20KB, provides a sensible compromise between memory requirements and recognition performance [ICAO:2004h, p. 31] & [ICAO:2004d, p. 36]. If, for the purpose of obtaining a digital version, a passport photograph with a face height of 25mm is to be scanned, the ICAO recommends a resolution of 300dpi [ICAO:2004h, p. 31].

Due to the varying compression rate of the JPEG-format though, the same compression factors can lead to quite different file sizes. In their study BioPII, the BSI used facial images with an average file size of 13,6KB [BioPII:2005, p. 30]. At the minimum, the file size amounted to 9,7KB, and the maximum size to 25,9KB. For storing iris images the ICAO recommends a file size of 30KB, and 10KB for fingerprints [ICAO:2004d, p. 32].

The BSI recommends higher values for iris images. In their study, the average size of a pair of iris images was 100,6KB [BioPII:2005, p. 34] while the average size of a pair of fingerprint images was 21,4KB [BioPII:2005, p. 32]. Based on these results, the ICAO recommends a minimum storage capacity of 32KB for RFID-chips, if only the facial image is stored [ICAO:2004d, p. 36]. For two additional fingerprints, at least 64KB are required. At present, two different chips are in use for the ePassports in current circulation: an Infineon SLE 66CLX641P with 64KB and a Philips Smart MX P5CT072 with a capacity of 72KB [BDR:2004, CCC:2005c].

4.3.7   Summary

The decision of the ICAO for the facial image as the first and solely binding biometric feature to be incorporated into the ePassport appears reasonable. This feature can be enroled for virtually everyone. The enrolment is relatively affordable. Moreover, the facial image in form of a passport photograph is already common, which in turn promises a high acceptance rate. However, the recognition rate is only moderate.

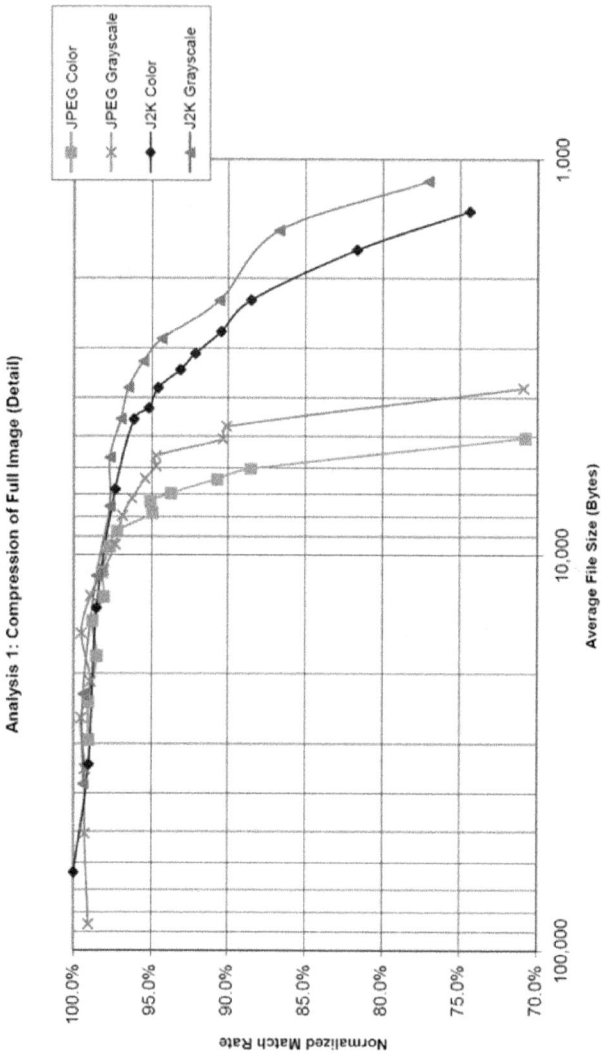

Figure 3: dependency between image size and the performance of biometric recognition

4.4 Security Features

4.4.1 Basic Access Control

As for the old non-electronic passport, by handing out the passport the holder indirectly agrees on the data contained in it to get read out.

In order to avoid any wireless unintended read-out of the data stored in the RFID-chip, the ePassport is provided with a special authentication mechanism, called Basic Access Control (BAC). BAC aims at enabling wireless data transmissions only after an optical reading device has read out specific data from the inside of the passport. This, in turn, requires the actual opening of the passport and close presence of the passport holder and therefore prevent unauthorized read outs.

The functionality of the Basic Access Control has been standardised by the ICAO and is described in the technical report [ICAO:2004a]. The following description is based on this particular ICAO document as well as on another document of the BSI [BSI:2005c].

First of all, the ePassport has to be opened in order to scan the Machine Readable Zone (MRZ) using Optical Character Recognition (OCR).

The data coded in the MRZ include the document number and some general facts on the passport holder, such as name, nationality, date of birth and gender [ICAO:2004a, p. 20]. Alternatively, in case OCR turns out to be impossible, this data can still be acquired manually, i. e. border police officers may themselves manually read-out the MRZ and type it into their computer for the actual verification process.

Similar to the computation of hash-values, the passport number, date of birth and expiration date are used for generating the access key that will subsequently be used for authenticating the reading unit towards the RFID-chip. Only these three MRZ elements are used, since they are the only ones being secured by checksums, which in turn allow errors occurring during the OCR-based read-out to be largely detected (and in general even corrected) automatically.

ICAO and BSI estimate the strength of the access key with at most 56Bit, which corresponds to the strength of an ordinary Single-DES key [BSI:2005c] & [ICAO:2004a, p. 56]. The ICAO also states that under certain circumstances, the strength of the key might be lessened. Section 5.5.3 will address this issue in detail.

The authentication process is carried out in several steps: At first, the chip generates a random number r_{Chip} and sends it to the reading unit. Afterwards, two ciphertexts are exchanged and encrypted with the access key K as described. Besides of the random number r_{Chip} received from the chip and a second random number r_{Reader} newly created by the reading device, the ciphertext being sent from the reading unit to the RFID-chip also consists of the first part K_{Reader} of the future session key. Using the access key K, the RFID-chip decrypts the ciphertext and compares the random number r_{Chip} it has sent to the reading unit previously, with the number contained in the ciphertext. In the positive case, a second ciphertext is being generated based on the access key K, consisting of the two random numbers r_{Chip} and r_{Reader} again, and the second part K_{Chip} of the session key. This ciphertext is sent back to the reading unit, which in turn decrypts the input and verifies the two contained random numbers. If they agree with the two sent to the chip in before, the two parts K_{Reader} and K_{Chip} are being joined to build the session key for the subsequent communication.

Figure 4 indicates the operational sequence of the Basic Access Control mechanism.

In the German version of the ePassport the subsequent communication between an RFID-chip and a reading device is encrypted with a 112Bit Triple-DES cipher [BSI:2005c].

However, since the access key with its strength of at most 56Bit is much weaker, a subsequent brute-force attack based on a complete recording appears at least theoretically possible.

The BSI nevertheless assesses the security level to be sufficient, since the Basic Access Control does not secure sensitive data like fingerprints, etc.

4.4.2   The Extended Access Control

In order to protect highly sensitive data like the fingerprint against unauthorised access, provision is made for the public-key authentication based Extended Access Control to be used from November 2007, in the course of launching the second stage of the ePassport.

This security concept, that is to extend the Basic Access Control, is still in the phase of specification at present. The mechanism relies on a session key negotiated via public key cryptography (Diffie-Hellman key exchange [DIHE:1976]). Furthermore, it provides the opportunity of a purpose limitation [Heise:2005e] & [BMI:2005d].

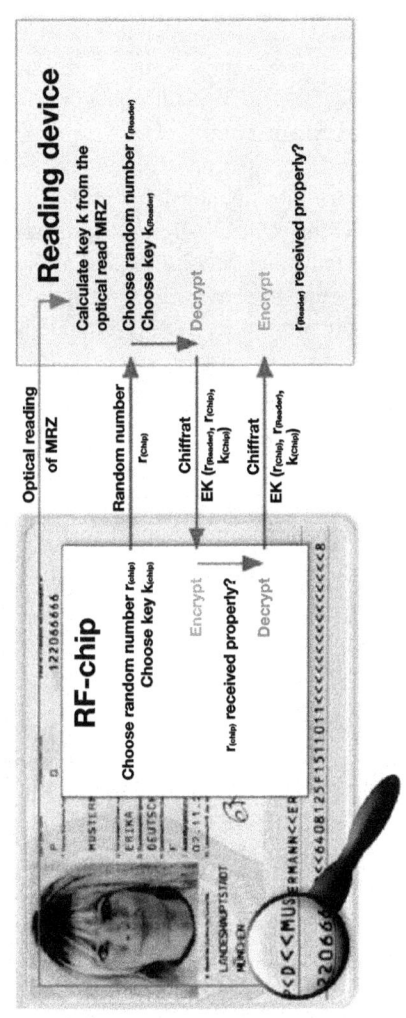

Figure 4: Basic Access Control (Source: [BSI:2005c, p. 3])

Only reading devices that feature a secret authentication key may be granted access to the data protected by the Extended Access Control. This way, the country issuing the ePassport is able to use digital signatures that determine which data may be accessed by which country in particular. This concept is explained in detail below. Concerning complete specifications, the reader is referred to [BSI:2006].

4.4.3 The Digital Signature

In order to ensure both the authenticity and integrity of the data, an ePassport is provided with a digital signature. Hence, it can be verified if the signed data was issued by a legitimate authority and if the data has been subsequently manipulated. The following description is based on two documents [BSI:2005c] & [ICAO:2004a].

All countries currently issuing ePassports are setting up a globally interoperable Public Key Infrastructure (PKI). In Germany, the Federal Office for Information Security (BSI) represents the central national authority for the key administration which for this purpose sets up a two-tier PKI, consisting of the so-called Country Signing CA (Certification Authority) and at least one Document Signer.

Country Signing CA: The Country Signing CA represents the supreme certification authority for each country. Its task is to certify the regional document signers.

Document Signer: The Document Signers are authorities entitled to issuing passport documents, for instance, the Bundesdruckerei in Germany. The Document Signer's private key is used to sign the information that is digitally stored in the ePassport (Document Security Objects (DSO)) and in doing so, protects them against manipulation. The Document Signer's key pairs have to be protected absolutely against unauthorised access. The private key is renewed on a regular basis to limit the losses in the potential case of compromisation. The Document Signer certificates are being submitted to the ICAO, which in turn is holding a Public Key Directory (PKD), in order to make the certificates available to all participating countries.

For both, the private key of the Country Signing CA as well as the private key of the Document Signer, a certain duration of usage is specified. This of course requires all signing authorities to renew their keys at regular intervals. In case of compromisation though, only certain documents then need to be questioned with respect to their authenticity.

The ICAO recommends a validity period between 3 to 5 years [ICAO:2004a] for the private key of the Country Signing CA and of not more than 3 months for the private key of the document signer. Based on a ten-year duration of validity for the German ePassport, this implies for the Country Signing CA to provide public keys with a validity period between 13 years and 3 months on one hand, and 15 years and 3 months on the other hand, depending on the actual duration of usage of the particular private keys they are based on. Correspondingly, the public key of the Document Signer is to be granted a validity period of 10 years and 3 months. For the case of compromisation, the ICAO intends the use of so-called Certificate Revocation Lists (CRL), that will be published on a regular base, enabling all countries in principle to identify potentially manipulated signatures [ICAO:2004a].

In Germany, the cipher ECDSA (Elliptic Curve Digital Signature Algorithm) will be used for the digital signing of the ePassports. For the Country Signing CA, the ICAO recommends a key length of 256Bit and 224Bit for the key of the Document Signer. Besides ECDSA, RSA and DSA are also accepted as ciphers.

4.5 Conclusion

The decision of using an RFID-chip in the ePassport appears seems reasonable. With the exception of the (contact-based) smart card, no other alternative provides a comparable storage capacity and in particular, has the opportunity of using a microprocessor to implement active security measures. The smart card has the disadvantage compared to the RFID technology that its contacts would not withstand the stresses and strains of a ten-year lifetime. A reflection on the durability of RFID-chips is done separately in Section 5.2

The mark left by biometry is ambivalent. Recognition performance has improved over recent years, and further improvements can be expected for the future. However, FRRs of 2–10% for face recognition mean that between 2 and 10 persons out of 100 may not be verified. A further review examination on biometrics is given in Section 5.2.2. The security measures for ensuring data privacy seem at first sight convincing. A closer inspection is done in Chapter 5.

5 A Critical View on the ePass

5.1 Introduction

The previous chapters described the ePassport with respect to its functionality in the way it was planned and finally implemented in Germany. Motivated by the criticism on the ePassport expressed by data privacy guards and security experts, this chapter examines further aspects that seem to be relevant for the overall evaluation. The following section covers the reliability of the system in general. The main focus thereby lies on the reliability of biometrics and on the durability of the RFID-chip. Moreover, possible attack scenarios performed by single individuals will be discussed. Ways for bypassing the system will be discussed in Section 5.4. The fifth section addresses data privacy and finally the sixth section deals with some further aspects mainly regarding political problems.

5.2 Dependability of the System in General

5.2.1 Introduction

As for the previous non-electronic passport, an issued ePassport is granted with a ten-year period of validity [AA:2005a]. Whereas for the old passport only the document itself (by means of material and binding) and the passport photograph were to last for a period of 10 years, this requirement now additionally holds for the RFID-chip together with the biometric data stored therein. More precisely, a ten-year usability of the ePassport demands for two things from the overall design. First, the physical and technical properties of material and electronic components must permit storage and reading-out of the data over the whole period of time. And secondly, the involved biometric features need to be suitable enough to allow reliable authentication up to ten years. It is still a controversial issue if in particular this second requirement is actually indeed the case. Below, the individual biometric features are being discussed with respect to their long-time use and dependability within the ePassport.

5.2.2 Dependability of Biometrics

As mentioned in Section 4.3, different studies report quite different results concerning the recognition rates of biometric systems. The BioPII study of the BSI examined the recognition performance for the features face, fingerprint and iris, and their suitability for being used within documents of identification [BioPII:2005]. The study concludes that *"biometric methods [...] are capable of effectively facilitating support for ID documents based verification of identity"* [10] [BioPII:2005, p. 169].

They further note that in practice, better recognition rates are to be expected, since *"the user will naturally have an immediate interest in the success of the verification process"* [10], and will therefore most likely follow the instructions and advice more willingly than participants of a study [BioPII:2005, p. 164].

This statement has to be treated with care. Actually, the test population, consisting of the staff of Frankfurt Airport, is not a true representation of the German population [BioPII:2005, p. 10]. Indeed, there are indications that the recognition performance within the context of a controlled operation might turn out to be less than the study suggests. From pages 51 et seqq. of the BioPII study, one can infer that the test population consisted ofa disproportionately high percentage of young to middle-aged, white-collar European males.

People with these properties achieve relatively high recognition rates (cf. Section 4.3). Men's features, in general, show more distinctive minutiae and larger fingers than women, allowing the fingerprint sensor to recognize male fingerprints more easily than female ones. Moreover, white-collar workers (professionals) are rarely subject to disturbing factors pertaining to their hands, e. g. injuries or calloused skin.

In contrast, and in particular people with Asian origin, often exhibit small fingers with extra fine fingerprint lines that are not conducive to successful enrolment and authentication. Finally, elderly persons actually tend to have worse results on biometric identification systems (cf. Section 4.3). And physically and mentally disabled people were completely ignored by the BioPII. Those people usually achieve significantly lower recognition rates and fail considerably more often to get enroled [UKPS:2005].

Although the BioPII study comes to the conclusion of biometric features to definitely exhibit the potential of effective support in the context of border controls, it still recommends *"an in-depth study of operational reliability, recognition performance and resistance against attacks"* [10], prior to the actual introduction of the ePassport [BioPII:2005, p. 170].

---

[10]Translated from German.

In reply to a written request, the BSI explained that this statement is expressly not to be interpreted in that way, (that such an in-depth study is to be carried out prior to the introduction of the ePassport) but rather *"prior to the introduction of biometric systems at the border controls"*[11]. The introduction of biometric systems at border controls was initiated at the beginning of 2006, and will probably last till 2009 (cf. Chapter 3.2). A corresponding study on the part of the Federal Government, however, has not been carried out yet [Bund:2005].

The BioPII study further states that the potential negative effects due to the aging of biometric features have not been investigated thoroughly. Because of this, it may be difficult to predict if biometric features recorded today may still allow for reliable verification after a period of ten years [BioPII:2005, p. 170]. The ICAO seems to share this scepticism. They recommend limiting the validity period of ePassports to 5 years, since the development of biometrics is proceeding rapidly and the recognition performance is likely to be decreasing over the years due to the aging of the corresponding biometric features [ICAO:2004d, p. 47]. Despite the recommendation, for now the German ePassport is granted with a ten-year period of validity [AA:2005a].

All in all, it can be stated that, given the recent progress in the field of biometrics over recent years, there is hardly any doubt that sooner or later adequate recognition performance will be achieved.

The only thing that is unclear at the present time is whether the biometric features prove to be operative despite the effects of ageing.

5.2.3 Durability of the ePassport

In practice, the ePassport has to face four primary sources of strain: Stamping, buckling and bending, dust and dirt, and finally the general ageing process of the RFID-chip. The hitherto non-electronic passport was considered to be fairly robust [BSI:2004b, p. 91]. Unfortunately, there are no studies that deal with the durability of RFID-chips embedded in documents. At present, experts are at variance about whether the ePassport will actually withstand all external stresses for ten years [BSI:2004a, p. 73].

In practice, stamping should not prove to be any problem. The recommendations of the ICAO are actually quite flexible regarding the question of where to embed the RFID-chip into the ePassport [ICAO:2004d, p. 41]. First images of the German Bundesdruckerei

---

[11]Translated from German.

gave reason to assume that the inlay with RFID-chip and antenna would either be embedded into the outer cover or into the data page of the passport.

Just as the prototype ePassport of the preliminary demo system [BDR:2004], the current final implementation now follows the first of the two options [BMI:2007b, BMI:2007d].

It can generally be stated that, depending on its design, the ePassport may indeed be damaged due to buckling [BSI:2004a, p. 45]. Dipl.-Ing. Jan Krissler, employee of the Fraunhofer Institute in Berlin, expressed a similar concern in that *"frequent buckling [...] may definitely damage the interface between chip and antenna"* [12] [Krissler:2005]. The ICAO obviously shares the opinion of buckling and bending to represent a realistic mechanical hazard. Therefore, they recommend to shore up critical parts using some rigid material [ICAO:2004b, p. 21].

Since RFID-chips are commonly considered to be resistant against dust and dirt [BSI:2004a], this factor can be considered as unproblematic.

Regarding the general ageing process, it is not yet clear to what extent they will affect the operational reliability of the RFID-chip.

The chip itself embodies an electronic storage medium. In practice, data cannot be stored on this medium indefinitely.

In this regard, the ICAO predicts an expected durability of RFID-chips und Smartcards of between 2 and 3 years [ICAO:2004d, p. 47].

According to a press report by Phillips, the type of RFID-chips that are used for ePassports is considered to be *"extremely reliable"* and *"the data remains preserved for much longer, compared to industry-wide common standards"* [12] [Philips:2005b].

The statements of the ICAO are somewhat confusing. They have previously stated that it is uncertain whether RFID-chips will continue to operate fully reliably after a period of five to ten years [ICAO:2004d, p. 47]. They suggest to limit the period of validity of passports to 5 years. Elsewhere, the ICAO states:

"There are now estimated to be in excess of 100 million Contactless ICs in circulation which conform to the ISO standards. The inherent durability of the Contactless ICs specified here is not in question."

This statement was made without mentioning any concrete numbers on how long these chips now "definitely" bear up [ICAO:2004d, p. 7].

---

[12]Translated from German.

In another document, the ICAO notes that RFID-chips retain their data for at least 10 years, based on a assumed storage temperature of 25° Celsius [ICAO:2004b, p. 17]. Of course, in practice, a passport will not continuously be stored at 25° Celsius. After all, the question of the long-term durability of RFID-chips still remains open.

5.2.4   Conclusion

Recapitulating, it can be stated that stamping, or dust and dirt are not highly likely to represent any hazard for the ePassport. Moreover, if appropriate provisions are taken during production, the risk of damage due to buckling and bending should be insignificant.

In regard to the aging process, it remains unclear to what extent RFID-chips will indeed be able to reliably store their data and stay fully functional for up to ten years.

If it should one day turn out that the durability of the ePassport actually does not meet the requirements and defective RFID-chips begin to accumulate, the ePassport will not have been proved to provide any nameable benefit in security compared to the old passport. After all, there is no way for a border police officer to determine which chip may have been damaged by the effects of the ageing processes or other stresses associated with the document, or if the chip had been intentionally corrupted.

5.3   Disturbance of Controlled Operation by Single Individuals

5.3.1   Introduction

Since the topic "ePassport" is being discussed not just controversially, but increasingly emotionally and not objectively [HOF:2005], the principle possibility of single individuals attempting to disrupt the operation of the ePassport must not be neglected. If, for instance, the respective individuals were to form an "anti-ePassport Movement", even though fractional in numbers compared to the anti-atomic power movement, and if there was a way to damage ePassports from a distance using electro-magnetic waves, high costs and the loss of the alleged security gain would be the direct consequence. In the course of the following sections, possible attack scenarios are examined and evaluated.

5.3.2 Jammers & Blocker Tags

Jammers and blocker tags aim at disturbing or interrupting the communication between the reading device and the ePassport. This may indeed hamper or even completely block any readout of the data stored on the RFID-chip [BSI:2004a]. A closer examination of this scenario will be omitted here. Actually, if it turns out that technically experienced ePassport detractors successfully develop mobile jammers, any attack on this basis can easily be counteracted by simply shielding the reading devices in an appropriate manner. In this manner, undisturbed communication can be assured [BSI:2004a].

5.3.3 Destruction by External Forces

In principle, there are three different methods of destroying RFID-chips in a non-mechanical way[13]:

- The memory contents of the EEPROM can be erased by intense E- and/or B-fields.
- The RFID-chip can be destroyed by impressing adequate high voltage on both connector pins, to which the coil is connected.
- An electrostatic charge causes a "lightning strike" onto the chip surface, leading to the destruction of the RFID-chip.

The effort for developing a mobile transmitter that is able to destroy or at least erase RFID-chips of ePassports at a distance of several meters is considered to be extremely high and associated with a lot of technical problems.

According to the state-of-the art, such a device would necessarily have to be fairly huge, and the power supply would be exceedingly problematic (cf. Appendix A). A realization of one of the other two options would require the attacker to at least temporarily obtain direct physical contact to the ePassport.

5.3.4 Demonstrations and Acts of Sabotage

At least theoretically plausible, notice the following scenario: Opponents of the ePassport could deliberately cause a high load of additional costs, e. g. in the form of selective demonstrations or

---

[13] According to a statement by Dipl.-Ing. Peter Jacob, employee of the EMPA, Department "Zentrum für Zuverlässigkeitstechnik" (formerly ETH Zurich's Institute for Construction Materials Testing). The exact wording of Dipl.-Ing. Peter Jacob can be taken from Appendix A.

acts of sabotage, similar to the acts of atomic-energy opponents in recent years [Welt:2004]. The benefit of the ePassport would clearly be cast into doubt, if due to its introduction people had to suffer in any way whatsoever. After all, the introduction of the ePassport has originally and primarily been motivated by and for increased interior security [BMI:2005c].

In practice however, the likeliness of such actions appear to be vanishing because with the introduction of the ePassport in its first phase did not lead to such activities. The further implementation of additional biometric data has been publicly known for a long time already, and even though severe criticism is still being expressed, no movement to our knowledge is emerging that would give reason to assume that radical tendencies necessarily resort to violence.

5.3.5 Conclusion

In principle, scenarios are imaginable in which the controlled operation of the ePassport could be disrupted. These scenarios, however, appear unlikely and should not seriously be considered as arguments against the introduction of the ePassport.

5.4 Fraud and Bypassing the System

5.4.1 Introduction

In order to achieve the aims of the ePassport (cf. Chapter 3.3) it needs to be ensured that the system's security measures can by no means be overcome. Compared to the former passport, it may be interesting to what extent the new ePassport represents a real improvement, or in the worst case, a deterioration.

Since the ePassport is based on the former passport and completely inherits all security features (cf. Chapter 3.4), and since the RFID-chip with the biometric data offers additional security features that have been granted to this document, it can be assumed that the ePassport will provide at least the same level of security. However, regarding the frauding or bypassing of the system, the following scenarios are being taken into consideration.

5.4.2 Authentic ePassport Based on Faked Papers

According to [Ross:2005], the question arises, if there will be a way for obtaining an authentic ePassport based on faked documents. In

Germany, when reporting a loss or theft of an ID-card or passport, one may apply for new documents of identification by providing another proof of identity, e. g. a birth certificate or driver's license [SKBS:2005]. If, for some reason, no such documents at all exist anymore, the proof of identity can also be provided in the form of a witness [SKBS:2005].

It appears obvious that documents such as a birth certificate or a driver's licence may be faked much easier than a (biometric) ID-card or passport. And even "false witness" may be obtained with adequate effort.

Having said that, one should take into account that the registration office has effectively full access to all the data of the original application. It is therefore actually possible for the officer to perform a cross-check by comparing the applicant with the passport photo from the archive file[14]. Hence, a successful application for an authentic ePassport based on faked documents appears to be quite difficult.

5.4.3 Falsified Passports from Countries not Using the ePassport

It may in principle be possible to falsify passports of countries that are not using ePassports as yet. In this case, however, besides the restriction of possessing a passport itself, the respective entry regulations will apply in addition. Thus, e. g. a visa needs to be applied for, which in turn involve a registration of biometric features, at least in Germany [BMI:2002b].

Presumably in 2008, all passport controls from the non-EU member state Switzerland to its neighbour states of the European Union, are going to be abolished [SchwBloe:2005]. At present, the ePassport is not mandatory for Swiss nationals [Schweiz]. As long as this situation remains unchanged, the standing of the ePassport must at least in parts be put into question. A more elaborate examination of this question would actually go beyond the scope of this book, though.

In summary, it cannot be completely ruled out that non-electronic passports from other countries that do not incorporate biometric data may get falsified, therefore permitting illegal entry into e. g. Germany.

---

[14]According to a statement of the director of the local residents registration office Magdeburg, this is indeed the common procedure.

5.4.4 Entry via Poorly Controlled Borders

According to the Federal Border Guard, during the period of the first half of the year 2000, a total of 15,217 illegal immigrants have been seized. It is quite predictable that the actual figure is much higher. According to [BM:2005], about one million people are currently residing in Germany without possessing a valid residence permit.

It is obvious that the introduction of the ePassport notwithstanding, illegal immigrants may still be likely to enter Germany at poorly controlled borders. Concerning this matter, a study of the London School of Economics & Political Science concludes that it would be much more worthwhile for the state to save the expenses for biometric equipped documents of identification, and instead delegate these funds towards enforced border controls. The study claims that funds used in this manner would definitely afford more effective protection against illegal entries into the country [LSE:2005]. This study, though, refers to Great Britain and is furthermore regarded by the British Government as being erroneous [UK:2005].

It is a moot point how far the funds dedicated to the ePassport could have provided more effectively to other provisions against terrorism and illegal immigration. At least as far as Germany is concerned, no such studies have been conducted as to what extent the objectives pursued by the ePassport may or may not have been accomplished by enhancing or funding stronger border controls.

5.4.5 Altering the Data Stored on the Chip

Due to their WORM architecture (write once, read many), the RFID-chips implemented into the ePassports cannot be (re)written or modified after their fabrication and initial write [BSI:2005a]. As a consequence, simply altering the data stored on the chip is not feasible. With the current implementation, the RFID-chip is embedded into the outer cover of the ePassport in a way that will hardly allow a replacement of the chip without noticeably damaging the document.

The data on the chip are furthermore protected by a digital signature of 224 of the respective document signer (cf. Section 4.4.3). Weak points in the architecture, however, might endanger this security mechanism. For instance, regarding the hash algorithm SHA-1, which is also being used for the ePassports' digital signatures, weak points are known for some time already. They allow reducing the complexity of an attack from an initial 2^{80} (brute force) to a 2^{69}

[WYY:2005] and recently even to a 2^{63} [Schneier:2005]. There is great reason to assume that with future research, the complexity may be reduced even further [Schneier:2005]. This security vulnerability, though, does not directly concern the ePassport. It takes effect only in case one tries to detect a collision between two random numbers or images. For the ePassport, however, it is relevant to find or detect a collision to a specific number or specific image. The aforementioned security vulnerability merely illustrates, after all, that it is difficult to predict how secure a cryptographic algorithm will be in the next years. The ICAO therefore recommends all countries presently granting passports with a ten-year period of validity, to reduce this validity to a period of five years only. This way, there will be a much more flexible response towards eventual progress on attacks on the algorithms [ICAO:2004d, p. 47].

5.4.6 Cloning an ePassport

Exact duplication of an ePassport may play a role in the extremely rare case of monozygotic twins trying to use the identity of their counterpart. The reason for this is because as long as the facial image is the only biometric feature being stored in the ePassport, the identical twins may have the continued ability of travelling with each other's passports. Provided that the required technical know-how is at hand, such a compilation of a 1:1 duplicate is in principle possible [ICAO:2004a, p. 17+55]. This technical possibility results from the fact that due to data privacy concerns, the RFID-chip embedded in the ePassport is not provided with a global unique serial ID. Consequently, RFID-chips as they are presently used in ePassports, have already been successfully cloned in practice [Heise:2006].

However, as soon as also fingerprints become part of the information stored, the ePassport enables an unambiguous identification, since even monozygotic twins feature different fingerprints [Philips:2005a]. Admittedly, for fingerprints as well as facial images, there is still the chance, though infinitesimal, of something called "Biometric Twins". However, the likelihood of biometric equity appears to be pretty much negligible [BSI:2003, p. 7].

Even though cloning of RFID-chips seems to be hardly relevant in practical terms, it still bears some risks. In fact, it would, for example, be possible for an attacker to create a manipulated clone of the RFID-chip in order to open up a communication with the reading device. Once this connection was established, the manipulated cloned chip could send malicious data that makes the reading-device crash or be infiltrated with a foreign code [Golem:2006].

Security experts like Lukas Grunwald, who was the first one to successfully clone the RFID-chip of an ePassport, regards cloning to be one of the biggest risks [Heise:2007]. Also, the German computer journal *iX* concludes after an in-depth analysis of the ePassport, that the complex architecture built on top of proprietary, yet not substantially tested standards, gives rise to a high vulnerability due to malicious software. [iX:2006, iX:2006b]

5.4.7 Fraud Resistance of the Biometric Verification

The security of the ePassport essentially depends on the fraud resistance of biometric verification. If it was possible to deceive, for instance, face recognition with an appropriately rouged face, a fingerprint recognition system with a silicone finger, or a iris recognition system with an appropriate contact lens, the whole benefit of the ePassport could well be questioned.

Studies conducted show rather negative results. [TKZ:2002] elaborates how all the established biometric systems can indeed successfully be deceived. Obviously, even fingerprint recognition systems with vitality detection may be deceived [MMYH:2002]. The German Chaos Computer Club further demonstrates how fingerprint recognition systems can easily be circumvented by fairly simple means [CCC:2004].

Latest studies report poor results of the biometric systems with respect to safety against overcoming. The study BioPI of the BSI concludes that the *"involved biometric systems can be overcome with only medium effort by creating copies of the biometric feature face in the form of a photo"* [BSI:2004b, p. 11].

The follow-up study BioPII gives reason to expect similar results for the fingerprint and iris. Besides primary test objectives, also the safety against overcoming of the involved systems was being checked in laboratories of the secunet AG [BioPII:2005, p. 11]. Unfortunately, apart from an overview on final grades, detailed results of these tests have not been published. On page 161 of the study, however, the given table contains a note that 3 of the 4 test systems have been assessed with grade "4" concerning their safety against overcoming, where "1" was the best and "6" the worst achievable grade. Grade "4" was given to those systems for which *"overcoming was successful with only medium effort (with granted access to the biometric feature of a legitimate person)"* [BioPII:2005, p. 158].

One system was granted a grade of "2" – but only thanks to the fact that vitality detection was applied. However, this system actually achieved even worse results regarding the "verification of a test person", which is not surprising, since vitality detection essentially leads to inherently higher false rejection rates [BioPII:2005, p. 63]. Moreover, vitality detection does not guarantee absolute protection. In [TKZ:2002], for instance, the vitality detection of a face was bypassed by something as simple as holding a water-filled bag in front of a passport photograph.

It remains open as to what extent provisions will be taken to prevent frauding and bypassing of the systems. Clearly, approaches as obvious as using photographs or even water-filled bags appear hardly feasible, since passport inspections will not be performed completely automatically, but will rather still be performed in the presence of border police officers. However, in order to detect silicone fingers or similar attempts, a much closer inspection of the travellers will be necessary. To what extent this will indeed take place, remains to be seen.

The BioPII study concludes that an in-depth study on the resistance against attacks preliminary to the final live operation is not only reasonable, but essential [BioPII:2005, p. 170].

5.4.8 Destruction of the ePassport's RFID-chip by the Passport Holder

In the event the RFID-chip of the ePassport gets destroyed, whether on purpose or not, and the biometric data are rendered impossible to read out, the ePassport will nevertheless keep its validity [BSI:2005a]. This means that even if a verification of the biometric features cannot be performed, entering Germany is still possible.

This measure is intended to ensure that possible technical problems may not result into a complete rejection of entry. However, at the same time, it questions the whole benefit of the ePassport. After all, what is gained by the introduction of the ePassport if entry is still possible despite a non-functional RFID-chip, i. e. without any verification of the biometric data?

According to information provided by the BMI, in the event of a non-functional RFID-chip *"the identity will be verified the classical way, in which case however this would prompt for a rigorous in-depth inspection."*[15] [CCC:2005a]. How such an "in-depth inspection" may take place is not detailed any further, though. Of cource, a comparison of fingerprints will not be an option, since the fingerprints of the passport holder are only stored in the form of a digital version on the RFID-chip. In particular, they will not be

---

[15]Translated from German.

printed onto the paper sheet[16], nor will they be stored in a database (cf. Section 5.5.7).

It might be well worth considering to include fingerprints into the paper part of the passport. Another option would be to check the fingerprints of the person to be verified against the data of the competent register of residents. Yet, it is indeed questionable how far this approach may in effect be possible, or even permitted by law.

5.4.9 Garbling of Biometric Features

As with the destruction of the RFID-chip, a similar effect would be due to garbling of the relevant biometric features, i. e. face and fingers. In doing so, a verification of the respective person against the biometric data stored on the RFID-chip is rendered impossible. Hence, the same points of criticism apply as mentioned in the previous section.

In effect, an individual who does not allow face or fingerprint recognition to be used for verification would get inspected more closely during passport control. The question arises, which possibilities are provided to the border police officers in the event that face and/or fingers are essentially unidentifiable.

However, such situations have already occurred in the past. Whenever a border police officer was being confronted with a traveller exhibiting a basically unrecognisable face, it was impossible to accomplish verification of the traveller against the passport photo.

According to information from two border police officers at Berlin Schönefeld Airport[17], there are no strict guidelines whatsoever applying in this case. Basically, it is up to the judgement of the border police officer to find an appropriate solution.

5.4.10 Conclusion

It remains unclear whether the goals of security would have been achieved in a better way if the funds that were invested for the ePassport had been delegated elsewhere, or if other options were considered. In the final analysis, it stands to reason that the safety against overcoming the biometric systems is insufficient.

---

[16]According to a telephone information provided by Michael Dickopf, press spokesman of the BSI.

[17]In reply to a personal request.

At the same time, it should also be considered that in case of a non-automated border control, the involved border police officer should be able to recognise observable attempts of manipulation like garbled photographs or silicone fingers. It is unclear, though, whether the border police officers will indeed always be sufficiently watchful and whether the reading devices will provide will perform in such a way that allows free insight in order to recognize attempts of manipulations reliably.

The fact is, that despite a broken RFID-chip, the ePassport still keeps its validity and might finally lead to the ePassport not exceeding the non-electronic passport by any means with respect to security.

5.5 Ensuring Data Privacy

5.5.1 Introduction

Data privacy activists criticize the introduction of the ePassport as being *"constitutionally highly problematic"*[18], and consider data privacy to be highly endangered [Heise:2005a]. This section discusses the concerns expressed most commonly.

In doing so, the options will be distinguished between those aimed at a selective obtainment of a specific person's data, and those allowing a mass readout of several persons' ePassports.

Regarding the first case, it is to note that with suitable effort, facial image as well as fingerprints of a specific person may very well be obtained even without accessing the ePassport. Thus, the focus of the following considerations will be on the question to what extent a mass readout of several persons' data may be possible.

5.5.2 Unauthorised Physical Readout of the Data

The BSI writes that using a "Focused Ion Beam" allows to decompose the RFID-chip step by step into atomic slices, rendering a readout of the chip possible [BSI:2004a, p. 48]. Admittedly, for this purpose both fairly high technical effort as well as direct access to the ePassport will be necessary. It appears reasonable that the effort required for temporarily getting possession of a foreign ePassport and analysing it with the use of highly sophisticated and fairly complicated technical methods may very well exceed the effort for obtaining facial image and/or fingerprints of the particular person in other ways.

Moreover, the biometric data is stored on the chip in a encrypted manner.[19] Thus, even if the whole content of the RFID-chip were to be successfully extracted, there would still remain the problem of getting all the data decrypted.

5.5.3 Cryptographic Security of Basic Access Control

The key used for the Basic Access Control consists of the expiration date, the passport holder's date of birth and the passport number. This leads to a maximum key length of 56 Bit at first glance (cf.

---

[18]Translated from German.
[19]According to a telephone information provided by Mr. Unger, employee of the BSI.

Section 4.4.1). In reality, however, the key needs to be considered weaker than 56 Bit.

Depending on the aim being either to obtain a single individual's data or to perform a mass readout on a number of passports, the domains of these three parameters can be restricted to more or less extent, resulting in a shorter effective key length.

First of all, within the first 10 years after the launch of the ePassport, the number of distinct expiration dates will not amount to 365×10, but by the end of the i-th year[20] to $365 \times i$ only. Considering all weekends and bank holidays, for which issuing offices will be closed, the number of possibilities decreases even further. In Germany more precisely by 52 weekends at 2 days each, plus at least 5 bank holidays per year[21]. This observation reduces the factor from 365×10 to 253×10, or $253 \times i$ by the end of the i-th year.

Taking into account that young and elderly people travel rather less frequently (and hence are met less frequently at the border control), and that those under 16 years old do not possess an electronic passport at all, one may assume a probable age of between 16 and 65 years, which in turn reduces the number of possibilities contemplable for the date of birth from 100×365 to only 49×365 (about 10^4).

If the circumstances allow an estimation of a person's age plus or minus 5 years, the number of possibilities can also be reduced further to 10×365 (about 10^3). Now, if for some reason the date of birth is even known exactly (e.g. in the event that the individual is personally known to the attacker), the multiplicative factor for this parameter completely reduces to 1.

The passport number consists – at least in Germany – of 9 digits, resulting in 10^9 theoretical possibilities, in principle. A possibly existing 10th digit can be neglected, since it just represents a checksum digit. In case these numbers are generated fully, randomly, or due to a pattern not known to the attacker, the domain of this parameter cannot be delimited further. This reveals that the security of the whole key crucially depends on the passport number, which in turn with 10^9 possible values in the ideal case makes up a much greater factor than the expiration date (about 10^3 possibilities) and the date of birth (between 1 and 10^4 possible values).

---

[20] Always with respect to 1st November 2005, the day the ePassport was launched in Germany.

[21] Between 8 and 12 bank holidays in Germany, depending the actual federal state. At least three bank holidays a year are on weekends.

	number of issuing days per year[22]	date of birth known exactly	date of birth estimated (± 5 years)	date of birth restricted (16–65 years)	date of birth unknown (0–100 years)
1 year	365	38	50	52	53
after launch	256	37	49	52	53
2 years	365	39	51	53	54
after launch	256	38	50	53	54
5 years	365	40	52	54	55
after launch	256	40	52	54	55
10 years	365	41	53	55	56
after launch	256	41	53	55	56

Table 3: Effective key length in Bit of Basic Access Control (BAC) for unknown authority ID (i. e., 10^9 possible passport numbers). Values rounded downwards to next full bit (i. e., truncated).

However, if the passport numbers are being generated either consecutively or based on a well-known pattern, the number of possibilities may be delimited significantly. In the Netherlands, for instance, passport numbers are issued in a consecutive manner. As reported, this procedure allowed the Basic Access Control (BAC) to get successfully broken within two hours using a customary PC, admittedly though, under optimal conditions, being that the birth date was known and the passport was valid for 5 years [Heise:2005b].

Like in the Netherlands, the passport numbers in Germany also are not generated randomly [PassG:1986]. Each of the 6500 passport issuing offices is being assigned a unique ID. This 4-digit number makes up the first digits of the passport number. The 5 remaining digits are assigned consecutively to the passports issued.

Obviously, the already reduced key strength may be again reduced significantly in case the ID of the issuing authority is known. In fact, if the ID of the issuing institution is known, but the five-digit sequential number is still unknown, the number of possibilities for this parameter of the key nevertheless reduces from 10^9 to only 10^5.

---

[22]256 days per year when assuming that no passports will be issued on weekends and bank holidays, 365 days otherwise

	number of issuing days per year[23]	date of birth known exactly	date of birth estimated (±5 years)	date of birth restricted (16–65 years)	date of birth unknown (0–100 years)
1 year	365	25	36	39	40
after launch	256	24	36	38	39
2 years	365	26	37	40	41
after launch	256	25	37	39	40
5 years	365	27	39	41	42
after launch	256	26	38	41	42
10 years	365	28	40	42	43
after launch	256	27	39	42	43

Table 4: Effective key length in Bit of Basic Access Control (BAC) for known authority id (i. e., 10^5 possible passport numbers). Values rounded downwards to next full bit (i. e., truncated).

Tables 3 and 4 provide an overview on how strong the key for Basic Access Control may be regarded, depending on particular circumstances. In this regard, Table 3 assumes the optimum of 10^9 possible random passport numbers. Table 4, in turn, assumes a successful reduction to only 10^5 possibilities, accounting for the case in which the ID of the issuing authority is known to the attacker, but not the passport number.
If the given circumstances moreover allow for a rough estimate on the passport number, the key length may even be delimited further. So, considering the observation that the effective key length may very well diverge from the theoretical key length, the question arises if such weaker key strength may nevertheless be regarded as being sufficient to guarantee data protection.

The data protection officer of the German Federal State Hessen recommends a 56 Bit key for use

"with non-sensitive personal data, and in cases where, due to other reasons, high-effort attacks are unlikely (e. g., in closed networks). However, future security concerns are to be expected."[24] [DH:2003]

---

[23]256 days per year when assuming that no passports will be issued on weekends and bank holidays, 365 days otherwise

[24]Translated from German.

He suggests the use of a key with a length of 40 Bit in turn only as "protection against accidental notice" [25] and for

"use in case of non-sensitive personal data, when a targeted attack is generally unlikely." [25] [DH:2003].

However, this recommendation, which is also being shared by other data privacy activists and data security experts, does not directly refer to the ePassport. It stands to assume that the microprocessor of an RFID-chip of course offers by far not the same performance as a customary PC or a system specialized for en- and decryption, which in general allow much faster interaction during the context of a brute-force attack.

In spite of this, scientists from the United States have shown able to decrypt an RFID-chip with a 40 Bit key within only one hour [BGSJRS:2005]. They actually even state the guess that the time required could very well be cut down to a few minutes.

The test setup, however, does not conform to the situation that applies for the ePassport. Moreover, a "live brute force attack" would require the attacker to permanently stay for a longer time within immediate proximity of the ePassport (or obtain possession) for the whole time of the attack. And last, but not least: Even if in the long run that chips were to be made more powerful, thus allowing a brute force attack within a time slot of a couple of seconds, such an attack could easily be counteracted by simply introducing an artificial delay between query and response. While an appropriate delay (e. g. one second) should very well be acceptable within the scope of passport control, it would render it (practically) impossible to systematically test all possible combinations, since this would most likely require much more time than available.

However, if the attacker were able to record the whole communication between ePassport and reading device, they would have all the time needed to decrypt the record and extract all the biometric data that have been transmitted. However, this scenario appears unlikely.

After all, recording the whole communication between ePassport and reading device requires the attacker to stay within immediate proximity of a few meters distance to the ePassport. Since communication with the ePassport should actually only take place at the border control, a mass recording of several passports in a row appears hard to put into practice, as the scene of recording over a longer period would hardly stay unnoticed. Finally, any recording

---

[25]Translated from German.

of the communication between reading device and RFID-chip could easily be prevented by shielding the zone around the reading devices [BSI:2004a]. Admittedly, such plans for shielding are not intended at present[26].

Provided that no such shielding is implemented, the recording of only a single person's communication, in turn, is much easier to accomplish. However, in Germany only the facial image is being protected by Basic Access Control. An attacker having the ability to record the whole communication and to decrypt it afterwards, should with less effort be able to obtain a facial image of the person of interest in other ways.

5.5.4 Bypassing the Basic Access Control

Regarding the Basic Access Control, Prof. Dr. Andreas Pfitzmann, employee of the Technical University of Dresden, notes this mechanism to be giving cause for concern regarding data privacy laws [Pfitz:2005]. Even if the technical aspects were to be classified absolutely secure, there would still be many persons being granted access to the MRZ of the ePassport, allowing them to read out the facial image stored in the RFID-chip.

Citing examples of persons with access to the MRZ, and hence to the whole key, Pfitzmann recalls the issuing authority, employees of the Bundesdruckerei GmbH (Germany's Federal Printing Office) and border police officers involved during the process of passport control. The list also includes companies and organisations, (e. g. banks or mobile phone service provider) where individuals are required to identify against using passport, id-card or a copy of one of them [Pfitz:2005].

This criticism appears basically justified. However, it is unknown to what extent it can be regarded critical when a person with direct optical access to the MRZ may be able to read out the facial image stored in the ePassport once again later on. This could be assumed a critical factor, on the supposition that the "real" passport photo attached to the paper part of the ePassport and the one digitally stored on the RFID-chip do not exhibit relevant differences.

As examples for further risks Pfitzmann mentions the generation of position tracking profiles and personalised bombs. These risks are being considered separately in Section 5.5.8.

---

[26]According to a telephone information provided by Michael Dickopf, press spokesman of the BSI.

All in all, due to its design, the Basic Access Control may indeed be bypassed, once an attacker has obtained access to the MRZ. Within immediate proximity to the RFID-chip, any such person will henceforth be able to read out the facial image as well as personal data stored on the RFID-chip of the ePassport without the passport holder noticing it. However, there are actually no additional data to obtain, that could not already have been acquired during the first access to the MRZ.

5.5.5 Cryptographic Security of Extended Access Control

Latest asymmetric ciphering methods such as the ECDSA, which is being used in the German ePassport, are basically regarded as secure [BleuKrueg:2001, p. 3]. However, since the ICAO has yet neither announced a suggestion nor published a standard for Extended Access Control, there is no basis for conducting a detailed in-depth analysis on its effective cryptographic security. A general consideration on this topic has already been conducted in Section 4.4.2.

5.5.6 Bypassing the Extended Access Control

Even without profound knowledge on how Extended Access Control works in detail, it can be stated that a security risk arises as soon as unreliable countries participate and are provided the necessary access keys. They would be given the ability to access all data released for their keys to the data stored on the RFID-chip, and all other countries would run into the risk of former ones passing their access keys to third parties or misusing the keys by themselves.

However, no country is obligated in any way to support other countries with the necessary access keys for Extended Access Control. According to the recommendation of the ICAO, only facial image access is to be granted to every country [ICAO:2004a]. Hence, apart from the facial image (which is already being treated by the Basic Access Control) it is up to Germany whom to grant access to which particular data, based on appropriate keys for the Extended Access Control.

5.5.7 Centralised Storage of Data

At present, any centralised storage of private personal data is clearly forbidden by law in Germany [PassG:1986, §4 (4)]. And in the direct run-up to the introduction of the ePassport, the German

Federal Government did not aim for changing the legal situation [Bund:2005]. However, in the near future, all passport photographs are going to be stored in a central database, from where they will be retrievable for police authorities [Heise:2007b]. Some parties additionally called for storing fingerprints as well, but this shall not be put into practice for the time being.

An in-depth discussion of the advantages and disadvantages shall not take place at this point. However, at the bottom line, it casts a damning light on the German Federal Government that prior to the ePassport's introduction any centralised storage of the data had been ruled out, just to finally indeed end up having them stored.

It should be pointed out that the issue of centralised databases is actually not relevant to the passport issuing country only. In this regard, the theoretical possibility of other countries establishing databases with biometric data of German travellers entering these countries will be considered in section 5.6.5.

5.5.8 Position Tracking & Personalised Bombs

Concerning other possible risks of the ePassport, Pfitzmann points out the possibility of position tracking and personalised bombs[27] being enabled due to the vulnerability of Basic Access Control (cf. Chapters 5.5.4 and 5.5.6) [Pfitz:2005].

Even though these aforementioned weak points exist, at least the generation of position tracking profiles does not appear feasible in practice. Even under optimal conditions, the operating range of the RFID-chip amounts to a maximum of a few meters only [FK:2004], and there is no reason to assume that in the long run reading devices will be installed at places other than border controls, and they will certainly not be installed with an area-wide deployment, which is a prerequisite for the creation of position tracking profiles.

Nevertheless, even in the most unlikely case that reading devices were installed in wide areas of Germany, there would be no way for the reading device to determine the person(s) in range.

The RFID-chip embedded in the ePassport does not exhibit a unique serial number. Hence, without having accessed the MRZ there would be no other way for the reading device but to try all possible MRZ's ever issued in Germany and to broadcast each of them to all ePassports within operating range. Only in the event of coincidentally having found a matching one, would the reading device

---

[27]The term personalised bomb refers to a bomb that automatically detonates as soon as a particular person enters a specific circuit around the bomb.

be able to identify the passport holder. All in all, a nation-wide individual position tracking seems out of the question, and more so since there are much easier ways for creating position tracking profiles, for example based on GSM mobile phone networks [BSI:2003].

The creation of personalised bombs appears, in theory, technically possible. As soon as an attacker acquires possession of the MRZ or of a valid key for Extended Access Control, profound technological knowledge will enable the attacker to construct a system that initiates a particular action, e.g. the detonation of a bomb, as soon as the corresponding ePassport is close enough to initiate the communication.

5.5.9 Improving Data Privacy

Most of the arguments against the ePassport mentioned in the previous sections result from the fact that it cannot be completely ruled out that unnoticed access to the data stored in the ePassport is possible under some conditions.

There would be less controversy if data privacy could be improved by making unnoticed readout more difficult or even impossible. We see the following starting points applicative:

The strength of the key used for Basic Access Control could be increased significantly if a "real" random key was to be used instead of a key that is composed of parameters that might possibly get delimitated significantly. (cf. Section 5.5.3).

The key used for Basic Access Control (as far as the concrete implementation of the ePassport is concerned, the MRZ) could be implemented in a way that only UV light would cause it to be unveiled. Then there would be no such problem as the key to be visible also on a regular photocopy, which is usually requested by mobile phone service providers or banks (cf. Section 5.5.4).

The ICAO mentions the means of incorporating a metal foil into the ePassport's cover. This would prevent the reading out of data from a closed passport [ICAO:2004i, p. 20] & [ICAO:2004b, p. 14+25]).

5.5.10 Conclusion

The facial image and further personal information like *Name* and *Date of Birth* are protected by the Basic Access Control; fingerprints in turn, are protected by the Extended Access Control. On closer examination, the Basic Access Control turns out to be not as secure as the BMI and BSI claim [BMI:2005d] & [BSI:2005c].

Admittedly, protection may still be sufficient, however, it is beyond reason why basically obvious mechanisms like the ones mentioned in Section 5.5.9 have not been implemented. An approach as simple as the incorporation of a metal foil into the cover of the ePassport would most likely have avoided a lot of the data privacy related provisos now being brought forward. In the meantime, such metal foils have even been put on offer already[28].

5.6 Further Aspects

5.6.1 Introduction

Besides data privacy related concerns and scepticism as to the reliability of the ePassport, critics find fault with further aspects, which shall be discussed in the following.

5.6.2 Indistinct Cost and Unknown Benefit

On 8th July 2005, the Federal Council of Germany was the last instance to approve the decision by the Federal Government regarding the introduction of the ePassport, however, without being given any concrete numbers on the potential costs for its introduction [BR:2005].

The estimated cost for the passport holder to obtain an electronic passport issued was fairly soon established and amounted to 59,- Euro. It is yet still unclear how much the actual costs, e. g. training the 35,000 employees, acquisition of the reading devices and equipping the 6,500 registration offices, will amount to finally [Bund:2005] & [Heise:2005c].

The Office of Technology Assessment at the German Parliament (TAB) predicts estimated extra expenses of between 180 and 610 million Euro plus additional annual costs of between 60 and 330 million Euro [TAB:2003, p. 142 et sqq.][29]. However, this is a decidedly rough estimation and is associated with several uncertainties [TAB:2003, p. 83].

---

[28]See, for example: http://www.pointprotect.de

[29]Note that this estimate, which was conducted in 2003, concerns the introduction of the electronic passport in general, and is not directly connected to the ePassport in particular.

The London School of Economics & Political Science performed an extensive study aimed at investigating the efficiency of biometric-based electronic ID-cards [LSE:2005]. They come to the conclusion that the introduction of biometric-based documents of identity would most likely indeed be effective against illegal immigrants and terrorists, but that these objectives could be achieved with other methods more efficiently [LSE:2005, p. 3].

In addition, they criticise the lack of reliability in biometric systems in general. The British government, in turn, expressly disagrees with this study [UK:2005].

Either way, the study can not directly be applied to German conditions, since it solely deals with the introduction of biometric id-cards, instead of passports, though with respect to similar aims.

As it is unclear how much the overall costs associated with the introduction of the ePassport will amount to in Germany, there is no basis for a well-founded cost-benefit analysis. An estimation as to how far the benefits will justify the expenses and to what extent all the objectives (cf. Chapter 3.3) of the ePassport may have been achievable in other ways, does not exist.

5.6.3 Overhasty Launch

According to data protection officers, the overhasty launch of the ePassport has lead to several problems [BDS:2005]. They note that the necessary prerequisites for a smooth launch are not yet sufficiently given.

In its decision on the introduction of the ePassport, the German Bundesrat complains about the Federal States having been *"included into the previous procedure only far too late and only insufficiently by the German Federal Government"*[30] [BR:2005].

Also members of the German Bundestag like Ulla Burchardt (SPD) criticise the resolution that has been passed *"based on a questionable basis and [...] despite open questions of technical, legal, and financial nature"*[30] (cf. Appendix C). Already at European level, the different countries' representatives are said to having been "completely run over" [Heise:2005a]. Some representatives in Brussels are said to have mentioned "blackmail" and a "perfidious game", and a British parliamentarian expressed her protest in that it was *"an absolute scandal that this attack on our civil liberties has gone over without any parliamentary examination whatsoever"*[30] [Krempl:2005].

---

[30]Translated from German.

Another aspect that has been criticised is that no case-studies or field tests have been conducted prior to the final decision of the ePassport's introduction, which in turn might have revealed obvious associated risks or allowed conclusions to be drawn regarding the positive aspects or the benefits regarding security to be expected [Bund:2005].

Furthermore, the resolution was being passed by the Federal Council of Germany at a time when studies of the BSI were already at hand. These studies assessed the performance of face recognition systems to be essentially *"at best sufficient in an automated scenario"*[31], but *"not acceptable"* in any way for everything beyond that [Bioface:2003, p. 10].

In the short run, it is still unclear to which extent the ePassport will actually be able to meet all the requirements arising from practice. During the first weeks, about 80% of the passport applications were rejected because the photos did not comply with the guidelines [Nehmzow:2005]. In most cases, the problem simply resulted from a lack of professional training of the photographers, which is presently not an issue anymore. However, there are still a few cases where photos did not comply with the guidelines just because head or nose shape of the applicant did not fit into the norm [Heise:2005i]. Although each of these applicants was finally issued a new passport, it remains questionable if the document will prove suitable for biometric authentication of its holder. Even more severe problems were revealed by the first ePassport Interoperability Test from 29th May 2006 to 1st June 2006 in Berlin. 30.6% of all tested reading devices were unable to keep the communication alive for the required time span due to technical problems [MunSei:2006]. Moreover, in 31.48% of the cases the Basic Access Control could not be performed successfully [Schlueter:2006].

After all, the questionably quick launch in Germany was explained with the emphasized increased urgency and the demands by the United States of America, as well as the potential economical benefit (cf. Chapter 3.3). The EU resolution, however, stated that the middle of 2006 was to be the time passports are to be supplied with a digital version of the facial image – plenty of time hence that could have been used for planning the launch more soundly and for resolving the doubt of the German public. The economic reason may indeed apply, as long as the ePassport operates properly. If, in contrast, the opposite turns out to be the case, a notably respective loss for the German economy is to be reckoned (cf. Chapter 3.3).

---

[31]Translated from German.

Over and above, Germany turns out to be not the only country feeling called upon to play the role of pioneer, regarding the development of the ePassport [BMI:2005c].

Austria, for instance, considers itself a leader, since the competence centre of the German Infineon company is located there, and the RFID technique of the Dutch Philips group is being developed in the Austrian city of Gratkorn [DP:2005].

The German Bundesdruckerei would not have benefited less, if the introduction of the ePassport were to be put into practice somewhat later, in accordance with the EU resolution as designated by the EU resolution in the first line. After all, the ePassport's launch itself might indeed be justified by economic benefits; the early date of 1st November 2005 in turn not.

The guidelines issued by the USA can as well only conditionally be accepted as a reason. Besides that, the USA's demand for biometric passports was only from October 2006 [BioPII:2005, p. 7], and one might also have went the way of Switzerland. In contrast to Germany, Switzerland allows its citizens a choice between a traditional conventional passport or a biometric equipped one [Heise:2005h]. Thus, persons who intend to travel to the USA, for example, may apply for an ePassport, whilst others may keep their non-electronic passport.

Finally, high urgency with respect to increased security requirements for the ePassports can also hardly be accepted as an argument in favour for the hasty introduction.

Upgrading the border controls with the required reading devices started in 2006 and is expected to be finished by 2008 [BSI:2005a]. The hitherto non-electronic passports will remain valid, so that the last non-electronic passports will expire not before 2015. Hence, it will still take a couple of years until the extra gain in security due to the ePassport will be effective to its full extent.

5.6.4 Information Policy

The main official information pages on the ePassport [BMI:2005a-e], [BSI:2005a-c], [Bund:2005] convey the impression of a mature and risk-free technology. In this regard, the German Federal Ministry of the Interior refers to the system as " a technically perfect solution", that has been "sufficiently tested"[32] [CCC:2005a].

---

[32]Translated from German.

At the same time, however, a study of the Federal Office for Information Security (BSI) concludes *"that the influence of aging effects on recognition performance has not been sufficiently investigated yet"*[33] and that *"prior to live operation in a concrete scenario, an in-depth evaluation on operational reliability, recognition performance and safety against overcoming is both suggested and necessary"*[33]. [BioPII:2005, p. 18]

On one of its ePassport information pages, the BSI claims that a readout of the data stored on the ePassport may, if at all, only be possible within a range of at most 15cm:

"For the RFID-chip used in the ePassport, an active reading out from beyond this range might be facilitated up to a maximum of about 15cm, with the help of an increased field strength used by the reading device. However, operating ranges beyond this are not realistic, due to physical principles."[33] [BSI:2005a]

Another study, also accomplished by employees of the BSI but not in concrete relation to the ePassport, remains completely unmentioned. That study reveals passive sniffing to be definitely possible within a range of up to 2 meters without much ado [FK:2004].

Amongst other objectives, the BSI investigated in its BioPII study, the transfer rate between ePassport and reading devices (cf. also Section 5.4.7). Unfortunately, not all details of the study got finally published. The BMI justifies this decision with the argument that in the course of the non-published part, methods for bypassing and overcoming biometric control systems have been described and examined with respect to their chances of success [BMI:2005f].

This statement actually leads us to assume that the ePassport's security against overcoming seems not to be in the best case, like it is usually claimed. Moreover, it appears highly questionable that some kind of "Security through Obscurity" approach should constitute a real measure of security that is best suitable for a project like the ePassport.

5.6.5 Political Challenges

New challenges arise at the political level. Each country, for example, will have to regulate with their individual PKI which country to grant access to the data for the optional biometric features, like fingerprints etc

---

[33]Translated from German.

As long as it is not to deal with obviously unreliable countries, it will be fairly difficult for politics to deny access to a particular country, if it does not wish to risk political strains. Thus, the question arises as how to prevent other countries to handle the biometric data in a way not intended by the issuing country.

For example, that German travellers, upon entering the USA, will not get their fingerprints stored in a central database. This problem, however, does not in the first place have any relation to the ePassport, since the USA is very well able (and willing) to collect fingerprints from all the travellers entering either way, with or without the ePassport. Nevertheless, the ePassport would basically allow all efforts to collect the desired data to be drastically reduced, making processing of the data more attractive for other countries as well.

The incorporation of more personal data into the ePassport could give rise to further desires, for example in commerce. Only three months after the launch of the ePassport, a BMI-internal discussion was launched as to whether personal data of the upcoming biometric ID-card should be offered to companies for payment of a suitable fee [Silicon:2006]. Admittedly, the discussion was not about selling the biometric data, but only general data like name, date of birth and place of residence. In the meantime, the BMI has finally discarded this idea, since it was just "stretching the imagination" [SZ:2006]. But already the fact that such ideas have once emerged at all and have been pursued further, instead of being discarded immediately, demonstrates how important an all-embracing inviolable protection of the ePassport data is.

5.6.6   Conclusion

At the European level, as well as in the German Bundestag and Bundesrat, politicians feel that they have not been adequately involved in the decision making process that has been exercised regarding the ePassport.

The introduction of the ePassport was decided without any specific knowledge as to estimated future costs to be expected. Up to that point, no case-studies or field tests were at hand attesting the maturity of the ePassport or documenting a real benefit.

Last but not least, at the bottom line, the information policy of the German Government and its related institutions appears in some points questionable.

5.7 Conclusion

Looking back at the provisos against the ePassport discussed in the previous sections, some of them appear indeed applicable after critical review.

For example, it remains to be seen whether or not the RFID-chips used for the ePassport will finally prove durable enough to guarantee reliable storage of the relevant data for a term of 10 years and more.

The fact that an ePassport with defective RFID-chip keeps being a valid travel document, until its regular expiration date, raises the question whether the Federal German Government may actually itself be in doubt about the long-lasting durability of the ePassport. After all, if defective RFID-chips were not expected to occur very often, there would be nothing to be said against an exchange that is free of charge but would have to be accomplished within a particular term after the initial manifestation of the defect.

The False Rejection Rates (FAR) for face and fingerprint recognition currently reside in the range of some percent. The BSI refers to this as being sufficient for assistance during identity checking.

The safety of the biometric systems against overcoming should be regarded deficient, whereas it remains open to what extent border police officers will be duly watchful during passport control with regard to potential attempts of manipulation.

The Basic Access Control (BAC) turns out to be not as safe as generally stated. At least at the present time, however, the safety that is ensured may actually be sufficient for the intended purpose.

Nevertheless, it is questionable why the option of using a metal foil for shielding has been completely abandoned.

Actually, this might have prevented a lot of the data privacy related provisos, and might have led to increased acceptance on the part of the German public.

The way the introduction of the ePassport was accomplished, as well as the haste exhibited, has met severe criticism on both the European and national level.

It appears somewhat incomprehensible, that more time was not taken to discuss the most important fears and reservations of the critics. Moreover, case-studies and field tests lacked in showing the effectiveness of the ePassport and its satisfactory functional capability.

All being considered then, from 1st November 2005, only ePassports are being issued. Yet, the equipment of border control stations with the necessary biometric systems just started at the beginning of 2006 and will endure until 2008 (cf. Section 3.2).

Consequently, even if in practice the biometric systems finally turn out to be unreliable in their live operation, only very few travellers may actually be affected and competent authorities are able to react appropriately, by either repairing the biometric systems or, in the worst case, suspending automated biometric checks temporarily.

6 Summary

The former German passport was said to be one of the most forgery-proof authentication documents worldwide. Yet, forgeries of passports issued by other, also European, countries occurred more frequently. Hence, the request for a substantial increase in safety is comprehensible (cf. Chapter 2). Accordingly, the decision of the European Union to arrange for an introduction of the electronic passport, binding to all EU member states, is basically sensible (cf. Chapter 3).

In Germany, the new generation of passports has been launched on 1st November 2005 under the label "ePass". In the current running first stage, the ePass is equipped with an RFID-chip storing the document holder's facial image. In the second stage, beginning in November 2007, two additional fingerprints will be stored. The price for the ordinary travel document with ten-year validity has been raised from 26 to 59 EUR.

Considering the technical requirements for storing biometric features, active security mechanisms and the consequent requirements as to memory capacity and transmission speed, the choice in favour of using an RFID-chip sounds reasonable (cf. Chapter 4.2). Also the determination for choosing face and fingerprints as biometric features appear plausible based on the current state of knowledge. (cf. Chapter 4.3). The security mechanisms Basic Access Control and Extended Access Control show that the issue of data privacy was by all means considered in the course of the design phase (cf. Chapters 4.4.1 and 4.4.2). Besides biometrics the digital signature additionally contributes to the high level of safety against forgery (cf. Chapter 4.4.3).

As pointed out in Chapter 5, many concerns regarding the ePassport expressed in the past can actually be considered unsubstantiated. For instance, the generation of position tracking profiles can practically be ruled out (cf. Chapter 5.5.8). Likewise, an unauthorised mass read-out of multiple ePassports seems hardly possible (cf. Chapter 5.5.4). Mechanical influences like stamping and buckling probably will not affect the ePassport's durability significantly (cf. Chapter 5.2.3).

On the other hand, some points of criticism do indeed qualify. It is questionable whether the RFID-chip used for the ePassport will finally prove capable of storing the committed data for the required period of 10 years (cf. Chapter 5.2.3). Apart from that, the aging effects of the involved biometric features have been investigated only insufficiently. In this respect it is unclear if, based on biometric features recorded today, authentication of identity will be sufficiently reliable in 10 years (cf. Chapter 5.2.2).

Similarly, the recognition performance of today's biometric systems is still not ultimately clarified. The BSI study BioPII comes to the conclusion that *"biometric methods [...] provide effective support for ID document based verification of identity"* [34]. However, as pointed out by the book at hand, the results that will be obtained in practice might turn out better or significantly worse (cf. Chapter 5.2.2). Apart from that, the same study has shown that the safety against overcoming biometric features may by no means be taken for granted. The BioPII study recommends conducting *"a profound examination of operational reliability, recognition performance and safety against overcoming"* [34], prior to a final live operation of biometric systems at border controls. Such an investigation has not yet taken place though.

To some extent, also the chance of unexpected advances in the field of cryptanalysis is considered a potential risk (cf. Chapter 5.4.5).

Due to the mentioned uncertainties regarding

   a. durability of RFID-chips,

   b. recognition performance of today's and future biometric systems w. r. t. aging effects, and

   c. continued preservation of data safety and data privacy, despite advances in the field of cryptography,

the ICAO, on whose recommendations the ePassport has been developed, suggests a five-year period of validity. Nevertheless, Germany decided to stick to granting the new passports a ten-year period of validity in general (cf. Chapter 5.2.2).

It has been pointed out that the security mechanism *Basic Access Control* shows vulnerabilities (cf. Chapter 5.5.3). These may, under certain circumstances, result in the key strength of the access keys to get weakened from initially 56 Bit down to only 28 Bit, or even less. Moreover, the Basic Access Control can be bypassed once access to the paper part of the ePassport has taken place. Apart from border control officers, this may (for instance) also apply to certain employees of banks or mobile phone service providers, having been handed out a photocopy of the passport in the course of concluding a contract. Although it appears indeed fairly unlikely, an issue like the construction of personalised bombs may not be completely ruled out due to this reason.

---

[34]Translated from German.

Three ways have been pointed out on how to enhance data privacy (cf. Chapter 5.5.9):

- The key strength for Basic Access Control may be increased, if a real random key as a separate part in the MRZ were to be used, instead of one consisting of three factors that might possibly get narrowed down significantly.

- If the credentials for building the Basic Access Control key, i. e. the MRZ, were to be visible only in the presence of UV-light, there would be no such issue as to undesirably pass on sensitive data with simple photocopies in the way they are handed out to banks or mobile phone service providers.

- The ICAO mentions the possibility of incorporating a metal foil into the cover of the ePassport. Such a foil would effectively prevent any read-out on a closed passport.

The fact that an ePassport keeps its validity despite any defects in the RFID-chip might potentially result in the measure of security provided by the new ePassport to only barely exceed that one due to its predecessor (cf. Chapter 5.4.8). If it should turn out that a considerable amount of RFID-chips became subject to age-related defects or complete failures, border control officers would certainly be unable to reliably differentiate between those RFID-chips that are non-functional due to age reasons and those that have intentionally been corrupted or destroyed. Thus, a person determined to prohibit any use of the stored biometric data, may easily achieve success by simply corrupting the RFID-chip.

The ongoing criticism regarding the unknown financial costs, the uncertain benefit, and finally the overhasty launch appears to be justified (cf. Chapter 5.6). In fact, the introduction of the ePassport was resolved even though the associated costs were largely unknown. Concrete studies examining to what extent the ePassport will be able to achieve the intended aims did not exist either at that time. On both the European and German level, politicians of various parties level criticism. The German Bundesrat found fault with the fact that the Federal States have actually been *"included into the previous procedure too late and too little."* [35]

At the bottom line, the overall information policy of the Federal Government can generally be criticised (cf. Chapter 5.6.4). The official data sheets about the ePassport communicate the impression of a mature and risk-free technology. In this respect, the German

---

[35]Translated from German.

Federal Ministry of Interior, for instance, refers to the implementations as "*technically perfect solutions*" that are definitely "*sufficiently tested*"[36]. At the same time, however, a study of the BSI (German Federal Office for Information Security) concludes that "*the influence of any aging effects to recognition performance of biometric systems has not been sufficiently investigated yet*", and that "*prior to the live operation in a concrete application an in-depth examination on operational reliability, recognition performance and safety against overcoming is both reasonable and necessary*".

Furthermore, the BioPII study on the one hand states that the biometric systems should be examined further with respect to their resistance against attacks. On the other hand, their very own test results have still not been published yet.

On one of its official information pages the BSI writes that any read-out of the ePassport's data may, if at all, only be possible up to a distance of not more than 15cm:

"*For the RFID-chip used in the ePassport, an active reading out from beyond this range might be facilitated within up to a maximum of about 15cm, with the help of an increased field intensity used by the reading device. However, operating ranges beyond this are not realistic, due to physical principles.*"[36]

However, another study carried out by the BSI, remains completely unmentioned, which reveals passive sniffing to be indeed possible within a range of up to 2 meters.

All in all, there may be hardly any doubt that an electronic passport storing biometric features of the document holder will prove effective against passport forgery and identity abuse. However, the overall procedure of its introduction, as well as some details of the technical realisation, were rightly criticized by the public in the past. This in particular applies to the major weak point of Basic Access Control that has been identified; to ignoring the ICAO's suggestion of a five-year term of validity for the new passports; to the uncertainty whether present biometric systems are already suitable for practice; and finally to the information policy of the involved public authorities.

Nevertheless, a real "*hi-tech disaster*"[36], which the German Chaos Computer Club (CCC) is afraid of [CCC:2005a], is certainly not to be expected. The introduction of the ePassport is performed gradually. The issuing of the first new ePassports was launched on 1st November 2005. The equipping of the border control stations

---

[36]Translated from German.

with the necessary technical systems, however, did not start before 2006 and will presumably last until 2008. Hence, at the bottom line, initial defects in the system may actually affect a small minority of travellers only, allowing the responsible authorities to smoothly take appropriate measures.

References

A

[AA:2005a]   Auswärtiges Amt. Gibt es verschiedene deutsche Reisepässe, ist der vorläufige Reisepass für die Einreise in alle Länder gültig?, 2005.

[AA:2005b]   Auswärtiges Amt. Was ist ein Kinderausweis?, 2005.

[Aetna:2005]   Aetna InteliHealth Inc. Iridology, 2005.

[AOK:2005]   Allgemeine Ortskrankenkasse AOK. Irisdiagnose, 2005.

B

[BDR:2004]   Bundesdruckerei. Bundesdruckerei legt Tempo für internationale ePass-Entwicklung vor, 02.11. 2004.

[BDR:2005b]   Bundesdruckerei. Personalausweis/Reisepass: Sicherheitsmerkmale der Personalausweiskarte, 2005.

[BDS:2005]   Bundesbeauftragter für Datenschutz. Entschließung zwischen der 69. und 70. Konferenz der Datenschutzbeauftragten des Bundes und der Länder, 01.06. 2005.

[BDSG:2007]   Bundesbeauftragter für den Datenschutz und die Informationsfreiheit. BfDI-Info 1. Bundesdatenschutzgesetz. Text und Erläuterung. 13. Auflage, 09 2007.

[BGSJRS:2005]   Steve Bono, Mathew Green, Adam Stubblefield, Ari Juels, Avi Rubin, and Michael Szydl. Security Analysis of a Cryptograhpically-Enabled RFID Device. Technical report, The Johns Hopkins University Information Security Institute Baltimore, 28.01. 2005.

[Bioface:2003]   Bundesamt für Sicherheit in der Informationstechnik. Bioface – Vergleichende Untersuchung von Gesichterkennungssystemen, Öffentlicher Abschlussbericht BioFace I & II, Version 2.1, 06 2003.

[Biofinger:2004] Bundesamt für Sicherheit in der Informationstechnik. Studie: "Evaluierung biometrischer
Systeme Fingerabdrucktechnologien – BioFinger", Öffentlicher Abschlussbericht, Version 1.1,
06.08. 2004.

[BioPII:2005] Bundesamt für Sicherheit in der Informationstechnik. Untersuchung der Leistungs-
fähigkeit von biometrischen Verifikationssyste-
men – BioP II, 2005.

[BleuKrueg:2001] Gerrit Bleumer and Heinrich Krüger-Gebhard.
Sicherheit moderner Frankiersysteme, 2001.

[BM:2005] Berliner Morgenpost. Studie: In Berlin leben bis
zu 200.000 Illegale, 04.07. 2005.

[BMI:2002a] Bundesministerium des Innern. Innenpolitischer
Bericht 1998–2002, 18.04 2002. Dr. Bernd Heim-
büchel.

[BMI:2002b] Bundesministerium des Innern. Der 11. September 2001 und seine Folgen, 01.03. 2002.

[BMI:2005a] Bundesministerium des Innern. Weiterentwick-
lung der Fälschungssicherheit von Pässen und
Personalausweisen, 2005.

[BMI:2005b] Bundesministerium des Innern. Bundesinnen-
minister Otto Schily zur Einführung des ePass
und zur Biometrie, 2005.

[BMI:2005c] Bundesministerium des Innern. Bundesinnen-
minister Schily stellt den neuen Reisepass mit
biometrischen Merkmalen vor, 01.06. 2005.

[BMI:2005d] Bundesministerium des Innern. Hintergrundin-
formationen zum ePass: Technik und Sicherheit,
2005.

[BMI:2005e] Bundesministerium des Innern. Fragen und
Antworten zum ePass, 2005.

[BMI:2005f] Bundesministerium des Innern. Schily weist
die Kritik des Bundesdatenschutzbeauftragten
zurück, 2005.

[BMI:2007] Bundesministerium des Innern. Kommunen
testen Fingerabdrücke für den ePass der zweiten
Generation, 27.02. 2007.

[BMI:2007b]   Bundesministerium des Innern. Informations-
                film zum elektronischen Reisepass: Beschrei-
                bung, 2007.

[BMI:2007d]   Bundesministerium des Innern. Fragen und
                Antworten zum ePass, 2007.

[BR:2005]   Bundesrat. Drucksache 510/1/05, 08.07. 2005.

[Bromba:2005a] Bromba. Bioidentifikation, 17.09. 2005.

[Bromba:2005b] Bromba. Fingerabdruckerkennung, 23.07. 2005.

[BSI:2003]   Bundesamt für Sicherheit in der Information-
                stechnik. GSM-Mobilfunk, Gefährdungen und
                Sicherheitsmaßnahmen, 2003.

[BSI:2004a]   Bundesamt für Sicherheit in der Information-
                stechnik. *Risiken und Chancen des Einsatzes von
                RFID-Systemen.* 2004. ISBN: 3-922746-56-X.

[BSI:2004b]   Bundesamt für Sicherheit in der Information-
                stechnik. Untersuchung der Leistungsfähigkeit
                von Gesichtserkennungssystemen zum geplanten
                Einsatz in Lichtbilddokumenten – BioP I, 2004.

[BSI:2005a]   Bundesamt für Sicherheit in der Information-
                stechnik. Häufig gestellte Fragen, 2005.

[BSI:2005b]   Bundesamt für Sicherheit in der Information-
                stechnik. BSI gewährleistet technische Sicherheit
                des elektronischen Reisepasses, 2005.

[BSI:2005c]   Bundesamt für Sicherheit in der Information-
                stechnik. Digitale Sicherheitsmerkmale im elek-
                tronischen Reisepass, 01.06. 2005.

[BSI:2005d]   Bundesamt für Sicherheit in der Information-
                stechnik. Biometrie – Gesichtserkennung, 2005.

[BSI:2006]   Bundesamt für Sicherheit in der Information-
                stechnik. Advanced Security Mechanisms for Ma-
                chine Readable Travel Documents – Extended
                Access Control (EAC), version 1.01. Techni-
                cal Report Technical Guideline TR-03110, 02.11.
                2006.

[Bund:2005]   Deutsche Bundesregierung. Drucksache 15/4616: Antwort der Bundesregierung auf die Kleine Anfrage der Abgeordneten Gisela Oiltz, Ulrike Flach, Rainer Funke, weiterer Abgeordneter und der Fraktion der FDP (Drucksache 15/4457), Biometrische Daten in Ausweispapieren, 04.01. 2005.

[BVerfGE:1983]   Bundesverfassungsgericht. BVerfGE 65, 1 – Volkszählung vom 15. Dezember 1983, 1983.

C

[CCC:2004]   Chaos Computer Club e.V. Wie können Fingerabdrücke nachgebildet werden?, 09.10. 2004.

[CCC:2005a]   Chaos Computer Club e.V. Auskunft des Bundesinnenministeriums, 2005.

[CCC:2005c]   Chaos Computer Club e.V. Der ePass – ein Feldtest. 22. Chaos Communication Congress, 27.–30. December 2005 in Berlin, 12 2005.

D

[DH:2003]   Professor Dr. Michael Ronellenfitsch. Zweiunddreißigster Tätigkeitsbericht des Hessischen Datenschutzbeauftragten, 2003.

[DIHE:1976]   W. Diffie and M. E. Hellmann. New directions in cryptography. *IEEE Transactions on Information Theory*, IT-22(6):644–654, 11 1976.

[DP:2005]   Die Presse.com. Österreich Musterschüler beim "ePass", 2005.

[Duden:2005]   Bibliographisches Institut Mannheim. *Duden – Das große Fremdwörterbuch. Herkunft und Bedeutung der Fremdwörter*. Dudenverlag, Mannheim, 2005. ISBN: 3-411041-64-1.

E

[EU:2004]	Europäische Union. Verordnung (EG) Nr. 2252/2004 des R vom 13. Dezember 2004 über Normen für Sicherheitsmerkmale und biometrische Daten in von den Mitgliedsstaaten ausgestellten Pässen und Reisedokumenten, 2004. Amtsblatt der Europäischen Union vom 29.12.2004.
[EU:2005]	European Commision. Technical Report of Biometrics at the Frontiers – Assessing the Impact on Society. Technical report, 02 2005. EUR 21585 EN.

F

[FK:2004]	Thomas Finke and Harald Kelter. Abhörmöglichkeiten der Kommunikation zwischen Lesegerät und Transponder am Beispiel eines ISO 14443-Systems, 2004. Whitepaper des BSI.

G

[GES:2005]	Gesundheit.com. Iriserkennung am Flughafen, 2005.
[Golem:2006]	Golem. E-Pass geklont, 2006.

H

[Heise:2005a]	Heise Online. Datenschützer verschärft Kritik an E-Pässen, 2005.
[Heise:2005b]	Heise Online. What the Hack: Hacken zwischen Kultur und Kurzschluss, 2005.
[Heise:2005c]	Heise Online. Deutschland setzt internationale Standards bei Biometriepässen, 2005.
[Heise:2005e]	Heise Online. Biosig 2005: Ein Pass, der passt, 22.07. 2005. Detlef Borchers.

[Heise:2005f]   Heise Online. Infineon und Philips liefern Chips für deutsche Biometriepässe, 2005.

[Heise:2005h]   Heise Online. Technologie mit Augenmaß: Der biometrische Pass in der Schweiz, 02.09. 2005. Detlef Borchers.

[Heise:2005i]   Heise Online. Und wenn der Mensch ein Mensch ist: Holpriger Start für den ePass. heise online, 03.11. 2005. Detlef Borchers.

[Heise:2006]   Heise Online. Sicherheitsexperte führt Klonen von RFID-Reisepässen vor, 2006.

[Heise:2007]   Heise Online. Zurückrudern in sachen ePass, 30.01. 2007.

[Heise:2007b]   Heise Online. Heftiger Streit um automatischen Polizeizugriff auf biometrische Passdaten, 12.04. 2007.

[HOF:2005]   Heise Online Forum. Biometriepass soll 59 Euro kosten, 2005.

I

[ICAO:2004a]   International Civil Aviation Organisation (ICAO). Technical Report – PKI for Machine Readable Travel Documents offering ICC Read-Only Access, Version 1.1. Technical report, 11 2004.

[ICAO:2004b]   International Civil Aviation Organisation (ICAO). Annex I – Use of Contactless Integrated Circuits In Machine Readable Travel Documents, Version 4.0, 05.05. 2004.

[ICAO:2004c]   International Civil Aviation Organisation (ICAO). Annex A – Photograph Guidelines, 2004.

[ICAO:2004d]   International Civil Aviation Organisation (ICAO). Technical Report – Biometrics Deployment of Machine Readable Travel Documents, Version 2.0, 2004.

[ICAO:2004e]   International Civil Aviation Organisation (ICAO). Technical Report – Machine Readable Travel Documents Development of a Logical Data Structure – LDS for Optional Capacity Expansion Technologies, Revision 1.7, 18.05. 2004.

[ICAO:2004f]   International Civil Aviation Organisation (ICAO). Facial Image Optimal Storage Size Study #1, 2004.

[ICAO:2004g]   International Civil Aviation Organisation (ICAO). Facial Image Optimal Storage Size Study #1, 2004.

[ICAO:2004h]   International Civil Aviation Organisation (ICAO). Biometric Data Interchange Formats – Part 5: Face Image Data, 2004.

[ICAO:2004i]   International Civil Aviation Organisation (ICAO). Supplement 9303, Version: 2005-5 V3.0, 12.06. 2005.

[iX:2006]   iX. ePass birgt Sicherheitsrisiken, 11.10. 2006.

[iX:2006b]   iX. Sicherheitsbetrachtungen zum ePass. iX 11/2006, p. 147, 11 2006.

K

[Krempl:2005]   Stefan Krempl. Biometrie statt Demokratie. in c't 26/2004, S.54, 2004.

[Krissler:2005]   Dipl.-Ing. Jan Krissler. Am besten in Alufolie einpacken. in die tageszeitung, 16.06.2005, 2005.

L

[LSE:2005]   The London School of Economics & Political Science. The Identity Project, An assessment of the UK Identity Cards Bill & its implications, 2005.

M

[MMYH:2002]   Tsutomu Matsumoto, Hiroyuki Matsumoto, Koji Yamada, and Satoshi Hoshino. Impact of Artificial "Gummy" Fingers on Fingerprint Systems. SPIE Vol. #4677, Optical Security and Counterfeit Deterrence Techniques IV, Thursday-Friday 24-25 January 2002, 2002.

[MunSei:2006]   A. Munde and U. Seidel. E-Passport Interoperability Test Event – Preliminary Results. BSI & BKA, 2006.

N

[Nehmzow:2005] R. Nehmzow. Viele Gesichter ungeeignet für neue Reisepässe. Abendblatt.de, 2005.

[NIST:2002]   NIST. Standards for Biometric Accuracy, Tamper Resistance, and Interoperability, 13.11. 2002.

P

[PassG:1986]   Das Deutsche Passgesetz, 1986.

[Pfitz:2005]   Andreas Prof. Dr. Pfitzmann. Biometrie – wie einsetzen und wie keinesfalls?, 2005.

[Philips:2005a] Philips Electronics. Für eine schnelle und bequeme Passkontrolle, 2005.

[Philips:2005b] Philips Electronics Pressemitteilung. Deutsche Regierung verwendet für elektronische Reisepässe hochsicheren kontaktlosen Chip von Philips, 2005.

R

[RFID:2002]   Klaus Finkenzeller. RFID-Handbuch, 2002.

[Ross:2005]   Philip E. Ross. Passport to Nowhere. *IEEE Spectrum*, pages 54–55, 2005.

S

[SchimKVK05]   Sascha Schimke, Stefan Kiltz, Claus Vielhauer, and Ton Kalker. Security Analysis for Biometric Data in ID Documents. In Edward J. Delph III and Ping Wah Wong, editors, *Security, Steganography, and Watermarking of Multimedia Contents*, volume SPIE Vol. 5681 of *Proceedings of SPIE*, pages 474–485. SPIE, 2005.

[Schlueter:2006]   M. Schlueter. Detailed Report of e-passport conformity testing: Layer 6-7. secunet AG, 2006.

[Schneier:2005]   Bruce Schneier. New Cryptanalytic Results Against SHA-1, 17.08. 2005.

[SchwBloe:2005]   René Schwok and Stephan Bloetzer. Die Beziehungen zwischen der Schweiz und der EU. Aus Politik und Zeitgeschichte (APuZ 36/2005), Bundeszentrale für Politische Bildung, 2005.

[Schweiz]   Schweizerische Eidgenossenschaft. Der Schweizer Pass.

[Silicon:2006]   Silicon. Berlin plant angeblich den Verkauf von persönlichen Daten, 2006.

[SKBS:2005]   Stadt Karlsruhe Bürgerservice und Sicherheit. Reisepass, 2005.

[Spiegel:2001]   Der Spiegel. Ausgabe Nr. 44/2001 vom 29. Oktober 2001, 29.10. 2001.

[SZ:2006]   Süddeutsche Zeitung. Treibt der Bund Handel mit Ausweisdaten?, 2006.

T

[TAB:2003]   Büro für Technikfolgen-Abschätzung beim Deutschen Bundestag. Biometrie und Ausweisdokumente, Zweiter Sachstandsbericht, Arbeitsbericht 93, 2003.

[TKZ:2002]   Lisa Thalheim, Jan Krissler, and Peter-Michael Ziegler. Biometric Access Protection Devices and their Programs Put to the Test. In c't 11/2002, 11 2002.

U

[UK:2005]   Home Office United Kingdom. Home Office Response to The London School of Economics' ID Cards Cost Estimates & Alternative Blueprint, 2005.

[UKPS:2005]   UK Passport Service.   Biometrics Enrolment Trial, 05 2005.

[ULDSH:2003]   Unabhängiges Landeszentrum für Datenschutz Schleswig-Holstein.   Datenschutzrechtliche Anforderungen an den Einsatz biometrischer Verfahren in Ausweispapieren und bei ausländerrechtlichen Identitätsfeststellungen, 2003.

W

[Watson:2005]   Craig Watson, 17.08. 2005.

[WDR:2005]   West Deutscher Rundfunk.   Sicherheits-Check per Irisabtastung, 2005.

[Welt:2004]   Die Welt. Die Welt: Höhn ruft zum Kampf gegen Trittin, 01.04. 2004.

[WYY:2005]   X. Wang, Y. Yin, and H. Yu. Collision Search Attacks on SHA1, 2005.

Appendix

A   Destructing an RFID-Chip (Translated from German)

Below an e-mail by Dipl.-Ing. Peter Jacob, employee of the EMPA, Department "Zentrum für Zuverlässigkeitstechnik" (formerly ETH Zurich's Institute for Construction Materials Testing) on the issue of destructing an RFID-chip:

There are basically three different ways to destroy an RFID-chip:

1.) By impressing a sufficiently high voltage on the two connector pins, to which the inductor is connected .

2.) By a "lightning strike" onto the chip surface through the chip's passivation, caused by electrostatic charges.

3.) By erasing the memory contents of the EEPROM through the introduction of extremely strong E- and/or B-field or deliberate manipulation via writing/reading device.

Let me now go into more detail on these three points.

To 1.): *The RFID chip is, in the card or transponder case, connected to either an antenna inductor or a small dipole antenna (according to the frequency band used). This antenna serves for both, the data exchange with the writing/reading device via radio link as well as the power supply of the chip. Depending on the distance between RFID and reading device, the antenna inductor yields a varying supply voltage. Because of this, RFID chips exhibit a voltage stabilisation or limiting circuit (besides a rectification of the AC voltage provided by the antenna). This way the chips internal operating voltage is being limited to a fixed value of between 2 to 5 Volts (depending on the respective chip technology). In general, any excess voltage is being turned into heat by diode lines, comparable to the well-known Zener diodes. If the input voltage level exceeds the supported breakaway range, it will very soon lead to the destruction of the chip due to EOS (electronic overstress), e. g., ablating of the supply conductor path. Using ordinary reading devices, this critical voltage will not be reached. However, a deliberate destruction may very well be achieved, for example by approaching the transponder card to a suitably electrified "primary" inductor (with only a few turns not sure this is correct - do you mean coils?, but high AC voltage). For dipole antenna based transponders the same effect* *could be caused by intense electromagnetic fields. Such an effect may for instance eventuate unintentionally from strong spark coils. At EMPA, we have observed such effects, when during testing RFID chips a spark coil was activated. In fact, latter one apparently had caused voltage peaks that the chip was unable to withstand for an extended period of time (>1 min).*

__To 2.):__ Generally, microchips, and so also RFID chips, feature some protection structure integrated into the external interconnections (pins), protecting the chip against temporary, usually electrostatically caused voltage peaks of 2kV at maximum. However, the situation looks different for electrostatic discharges directly impacting onto the chip surface (surface ESD, ESDFOS). In this case the about 1μm thick chip passivation will be penetrated, and short circuits will be caused between the two upper metal conductor path layers. This fault mechanism is known as a "product killer" in the course of chips into plastic cases, cards, glass capsules, etc. In the housed state, the chip is well protected against electric impact. However, in case one were to deliberately cause destruction, then a targeted lightning strike penetrating the card and impacting onto the chip surface, for instance through the use of a Van-de-Graaf generator, influence machine or spark coil, would yield the desired effect and destroy of the RFID chip.

__To 3.):__ Given a suitable writing device with the necessary access codes, one could manipulate the chip by simply overwriting the data of the RFID internal EEPROM. An alteration of the EEPROM contents via intense electric and/or magnetic fields or UV or alpha radiation is, from a physical point of view is, in principle not impossible – however, in fact only if the magnetic fields or ratiation is extremely strong. Being aware of this possibility, RFID chips are being inspected in the scope of their eligibility regarding their resistance against this kind of affection. (There exists also a standardisation in this regard, which unfortunately I do not have at hand right now, but could probably search out if needed.) For magnetic fields, for instance, such checks are being conducted up to Tesla range, though I do not recall any failures for the RFID types designs currently known to me.

B   Destructing an RFID-Chip (Original Email in German)

Eine Email von Dipl.-Ing. Peter Jacob, Mitarbeiter der EMPA, Abteilung "Zentrum für Zuverlässigkeitstechnik" (ehemals das Institut für Baumaterialprüfung der ETH Zürich) zum Thema Zerstören von RF-Chips:

Eine Zerstörung eines RFID-Chips kann grundsätzlich auf drei Arten erfolgen:

1) Anlegen einer sehr hohen Spannung an die beiden Anschlusspins, an welcher die Spule angeschlossen ist

2) "Blitzeinschlag" in die Chipoberfläche durch die Chip- Passivierung hindurch infolge elektrostatischer Aufladungen

3) Löschung des EEPROM-Memoryinhalts durch Einbringen extrem starker E- und/oder B-Felder bzw. gezielte Manipulation über Schreib-/Lesegerät.

Nun zu den einzelnen Punkten im Detail:

zu 1. *Das RFID-Chip ist in der Karte oder im Transpondergehäuse mit einer Antennen-Spule oder einer kleinen Dipolantenne (je nach Frequenzbereich) verbunden. Über diese Antenne erfolgt sowohl der Informationsaustausch via Funkstrecke mit dem Lese-/ Schreibgerät als auch die Speisung des Chips. Je nach Abstand des RFID vom Lesegerät wird durch die Antenne eine sehr unterschiedliche Versorgungsspannung erreicht. Aus diesem Grund haben RFID Chips neben der integrierten Spannungsgleichrichtung der aus der Antenne zugeführten Wechselspannung auch eine Spannungs- Stabilisierungsschaltung oder zumindest -Begrenzung eingebaut. Dadurch wird die interne Betriebsspannung des Chips auf einen Festwert zwischen 2-5 Volt - je nach Chiptechnologie - begrenzt. Die überschüssige Spannung wird dabei meist durch begrenzende Diodenstrecken, etwa vergleichbar den bekannten Zenerdioden, in Wärme umgesetzt. Übersteigt die Eingangsspannung den möglichen Abregelbereich, so erfolgt nach kurzer Zeit eine Zerstörung des Chips infolge EOS (Electrical Overstress), z.B. durch Abschmelzen der Versorgungsleiterbahn. Mit normalen Lesegeräten wird diese kritische Spannung nicht erreicht. Eine gewollte Zerstörung könnte aber beispielsweise durch Annähern der Transponderkarte an eine entsprechend bestromte "Primär"-Induktionsspule (mit wenigen Windungen aber hoher Wechselspannung) erfolgen. Bei Dipolantennen-Transpondern würden überhöhte Elektromagnetische Felder des ent-*

*sprechenden Frequenzbereichs den gleichen Effekt bewirken. Ein sol-
cher Effekt kann zum Beispiel durch starke Funkeninduktoren unge-
wollt eintreten. An der Empa haben wir solche Effekte festgestellt,
als RFID-Chips elektrisch getestet wurden und ein Funkeninduktor
in der Nähe betätigt wurde. Durch diesen wurden in der Testzulei-
tung des Chips Spannungsspikes erzeugt, denen das Chip nicht über
längere Zeit (>1 Min) gewachsen war.*

*zu 2.   Generell haben Mikrochips, so auch RFID-Chips, an den
elektrischen Aussenverbindungen (Pins) Schutzstrukturen eingebaut,
die vor kurzzeitigen, meist elektrostatisch verursachten Spannungsspikes bis etwa maximal 2kV schützen. Anders verhält es sich aber
bei elektrostatischen Entladungen, welche direkt auf die Chip-Ober-
fläche einwirken (Oberflächen-ESD, ESDFOS). Dabei wird die etwa 1um dicke Chip-Passivierung durchschlagen und es entstehen
Kurzschlüsse zwischen den beiden oberen Metall- Leiterbahnebenen.
Dieser Fehlermechanismus ist als "Produktkiller" bei den Einge-
häusungsprozessen der Chips in Plastikgehäuse, Karten, Glaskap-
seln usw. bekannt. Im gehäusten Zustand ist der Chip hingegen gut
geschützt gegen einen direkten Einschlag. Würde man jedoch eine
bewusste Zerstörung herbeiführen wollen, so könnte etwa ein geziel-
ter Blitzeinschlag, welcher die Karte durchschlägt und die Chipoberfläche trifft, etwa mit Hilfe eines Van-de-Graaf-Generators, In-
fluenzmaschine oder Funkeninduktors oder dgl. die Zerstörung des
RFID-Chips bewirken.*

*zu 3.   Sofern ein entsprechendes Schreibgerät mit Zugangscodes
vorliegt, könnte durch Löschen/ Überschreiben der Daten des im
RFID befindlichen EEPROMS eine Manipulation des Chips ge-
macht werden. Eine Änderung des EEPROM-Inhalts durch starke elektrische und/ oder magnetische Felder sowie auch durch UV-
und radioaktive alpha- Strahlung ist physikalisch zwar grundsätzlich
möglich, aber nur bei Ansatz jeweils extrem starker Einwirkung. Im
Bewusstsein dieser prinzipiellen Möglichkeit werden RFID-Chips im
Rahmen ihrer Qualifikation auf ihre Resistenz gegen diese Einwir-
kungen mustergeprüft. (Dafür gibt es auch einen Standard, ich habe diesen allerdings leider gerade nicht zur Hand, könnte diesen
aber evtl. herausfinden.) Bei Magnetfeldern werden solche Prüfun-
gen beispielsweise bis in den Tesla-Bereich geführt, wobei es bei den
mir bekannten RFID-Baumustern bisher zu keinen Ausfällen ge-
kommen ist.*

C Email by Ulla Burchardt (SPD), Member of the German Bundestag (Translated from German)

Dear Mr. Beel,

Thank you very much for your e-mail from 14. September 2005 regarding the subject of biometrics. As you have correctly noted, several times already I have argued against the introduction of biometric features in passports.

My criticism can be outlined in a few sentences: According to applicable law, the German Bundestag should have been the authority to decide on the introduction of this technology. However, that never occurred , because a decision has been brought about on a European level by means of an EU regulation.

On a European level however, the so-called "Hearing Procedure" has been put into effect. As a consequence, the European Parliament simply did not have had any chance to enforce its various requests for change against the influential European Council of Ministers. Apart from that, the original draft of the regulation of the European Parliament's committee, has later been subject to serious modifications by the European Council of Ministers. Finally, the European Union may take action only if it exhibits clear authority and the biometric passport is still currently controversial.

In Conclusion, given such an important topic as the ePassport, parliaments have been left out. Instead, the decision was taken by the European Council of Ministers, grounded briefed on a questionable legal basis and despite still open technical, legal and financial issues.

If you have further questions, please don't hesitate to contact my Berlin office. I would also be interested to hear more about your project and the subject you are developing.

I wish you a lot of success in your future studies.

Kind regards,

sgd. Ulla Burchardt

D Email by Ulla Burchardt (SPD), Member of the German Bundestag (Original Email in German)

Sehr geehrter Herr Beel,

haben Sie vielen Dank für Ihre E-Mail vom 14. September 2005 zum Thema Biometrie. Wie Sie zutreffend schreiben, habe ich mich bereits mehrfach gegen die geplante Einführung biometrischer Merkmale in Pässen ausgesprochen.

Meine Kritik am Verfahren lässt sich in wenigen Sätzen zusammenfassen: Der Bundestag hätte nach geltendem Gesetz über die Einführung beschließen müssen, wozu es aber nicht kam, weil eine Beschlussfassung auf europäischer Ebene mittels einer EU-Verordnung herbeigeführt wurde.

Auf europäischer Ebene wiederum kam das so genannte "Anhörungsverfahren" zur Anwendung, das Europäische Parlament hatte also keine Möglichkeit, seine mannigfachen Änderungswünsche gegenüber dem allein maßgeblichen EU-Ministerrat durchzusetzen. Im Übrigen wurde der Verordnungsentwurf nach Abschluss der Beratungen im federführenden Ausschuss des Europäischen Parlaments vom Ministerrat noch gravierend abgeändert. Und schließlich: Die EU darf nur dann tätig werden, wenn sie eine ausdrückliche Kompetenz dafür hat. Genau das aber ist beim biometriegestützten Reisepass zumindest umstritten.

Fazit: Faktisch blieben die Parlamente bei einem so bedeutsamen Thema wie dem ePass außen vor, beschlossen hat der EU-Ministerrat auf einer fragwürdigen rechtlichen Grundlage und das trotz ungeklärter technischer, rechtlicher und finanzieller Fragen.

Gerne können Sie sich im Falle weiterer Fragen mit meinem Berliner Büro in Verbindung setzen. Im Übrigen würde ich mich freuen, wenn Sie mir das genaue Thema und die Fragestellung Ihrer Arbeit in einer kurzen E-Mail noch etwas näher erläutern würden.

Ich wünsche Ihnen auf Ihrem weiteren Studienweg viel Erfolg und verbleibe

mit freundlichen Grüßen

gez. Ulla Burchardt

www.ingramcontent.com/pod-product-compliance
Lightning Source LLC
Chambersburg PA
CBHW071722170526
45165CB00005B/2117